Resonance

Chakra Balance and the **Law of Attraction**

LUELLA GOETHALS

A DIVISION OF HAY HOUSE

Copyright © 2024 Luella Goethals.

All rights reserved. No part of this book may be used or reproduced by any means, graphic, electronic, or mechanical, including photocopying, recording, taping or by any information storage retrieval system without the written permission of the author except in the case of brief quotations embodied in critical articles and reviews.

Balboa Press books may be ordered through booksellers or by contacting:

Balboa Press
A Division of Hay House
1663 Liberty Drive
Bloomington, IN 47403
www.balboapress.com
844-682-1282

Because of the dynamic nature of the Internet, any web addresses or links contained in this book may have changed since publication and may no longer be valid. The views expressed in this work are solely those of the author and do not necessarily reflect the views of the publisher, and the publisher hereby disclaims any responsibility for them.

The author of this book does not dispense medical advice or prescribe the use of any technique as a form of treatment for physical, emotional, or medical problems without the advice of a physician, either directly or indirectly. The intent of the author is only to offer information of a general nature to help you in your quest for emotional and spiritual well-being. In the event you use any of the information in this book for yourself, which is your constitutional right, the author and the publisher assume no responsibility for your actions.

Any people depicted in stock imagery provided by Getty Images are models, and such images are being used for illustrative purposes only. Certain stock imagery © Getty Images.

Print information available on the last page.

ISBN: 979-8-7652-5507-0 (sc)
ISBN: 979-8-7652-5508-7 (e)

Balboa Press rev. date: 08/29/2024

Disclaimer Notice

The contents of this book, including any techniques, practices, and advice, are provided for informational and educational purposes only. The author is not responsible for any direct, indirect, consequential, special, exemplary, or other damages that may result, including but not limited to economic loss, injury, illness, or death. The practices discussed herein, such as chakra balancing, meditation, and the Law of Attraction, are not substitutes for professional medical advice, diagnosis, or treatment. Always seek the advice of your physician or other qualified health provider with any questions you may have regarding a medical condition. Never disregard professional medical advice or delay seeking it because of something you have read in this book.

The true stories shared in this book are based on the author's personal experiences or those of others who have given written permission for their inclusion. Any resemblance to actual persons, living or deceased, or actual events is purely coincidental. Certain details, including names, locations, and identifying characteristics, may have been changed to protect the privacy and confidentiality of the individuals involved. The perspectives and interpretations provided are those of the author and may not necessarily represent the views or experiences of the individuals involved.

The author makes no guarantees or warranties, express or implied, about the completeness, reliability, or accuracy of the information contained in this book. The reader assumes full responsibility for any actions taken based on the information provided within. The reader should not construe the stories as endorsements or guarantees of any particular outcome. The author shall not be held liable for any direct, indirect, incidental, consequential, or other damages resulting from the use or misuse of the information contained in these stories.

The quotes and excerpts used in this book from historical figures such as Nelson Mandela, are used in accordance with fair use principles. These quotes are public statements and are attributed accurately to their original sources.

The author's interpretations of principles like the Law of Attraction, Law of Vibration, and other metaphysical concepts are intended to be original thoughts and are not direct reproductions of any other author's work. However, references to other works are included where relevant and are acknowledged in the reference list provided.

By reading this book, you acknowledge that you have read, understood, and agreed to this disclaimer.

Contents

We are Travellers in Life 1
Concepts and Definitions 4
Illuminating Our Path: We Are Beacons of Light 9
The Subconscious Mind 13
    True Story; The Stranger 15
The Root Chakra: Resonance of Intention 18
    True Story: Uncle's Energy 23
The Sacral Chakra: Resonance of Empowering Emotions ... 27
    True Story: Martha's Fear 39
The Solar Plexus Chakra: Resonance of Inner Strength 41
    True Story: Bringing Light to Shadows 50
The Heart Chakra: Resonance of Healing 53
    True Story: The Gift 58
The Throat Chakra: Resonance of Expression 60
    True Story: Louise's Weekend 64
The Third Eye: Resonance of Insight 67
    True Story: Seeing Clearly 71
The Crown Chakra: Resonance of Divine Connection 73
    True Story: The Apple Tree 75

Testimonials 83
References 85
About the Author 87

We are Travellers in Life

> *"Life is a journey, not a destination."*
> – Ralph Waldo Emerson

OVER THE PAST TWO YEARS, AS BOTH A LIFE AND CAREER COACH, I have deepened both my conventional coaching and my intuitive sessions. During these intuitive encounters, I meditate and connect with my clients' chakra systems. Conducting over thirty sessions weekly has led me to a profound realisation: we are all travellers in life, always exactly where we need to be, for life is a journey rich with experiences. Embracing the present moment is key, as it is within this presence that life's true purpose unfolds—to simply be.

Each individual journey is unique, and no one else can define our path. As spiritual beings inhabiting physical forms, our body, mind, and soul all require nurturing. The 'soul' refers to the invisible essence within us, which is enriched and shaped by our life experiences on this earthly plane. Our challenges are not merely obstacles but powerful lessons that contribute to the growth and evolution of our soul.

Our ultimate goal is to resonate with vibrations that support growth—whether through happiness, serenity, or other nurturing energies. By understanding and applying the Law of Attraction, and by engaging the power of our chakras, we align our thoughts, emotions, and energies to magnetise the essential vibrations our souls seek, such as serenity or lightness. These vibrations are the light of the soul. The Law of Attraction elucidates how our thoughts and feelings sculpt our reality. Concentrating on our desires rather than our fears allows us to manifest our dreams into existence. Throughout this journey, we must embrace lessons of release and compassion, ensuring our energy flows outward, aligning our body, mind, and soul like a potent antenna.

Our chakras, subtle energy centres within our bodies, play an integral role in maintaining both our spiritual and physical health. By balancing and activating these chakras, we harmonise our energy, which leads to greater fulfilment. Each chakra corresponds to specific life aspects, and nurturing them enhances self-awareness and inner peace. This process, in turn, activates the Law of Attraction, an invisible but omnipresent force, epitomised by the principle that like attracts like. In states of serenity, miracles seem to unfold, and opportunities appear as though by magic—such is the enigma of the Universe.

So, how does one begin to infuse their life with serenity and beauty? It starts within you, akin to a tree analogy. Like trees, our root chakras need nourishment just as a tree's roots absorb water and nutrients from the soil. When nourished, we stand tall and strong, resembling a tree with a robust trunk. Our crown chakra, similar to the uppermost branches of a tree, soaks in sunlight, drawing in the energy that propels our growth.

In the face of adversity, just as a tree bends and sways to counterbalance the storm, we too learn to centre ourselves and adapt to challenges, maintaining balance amidst upheaval. This resilience and adaptability are what help us maintain our alignment and continue to thrive.

By embracing the Law of Attraction and nurturing our chakras, we embark on a profound journey of self-discovery and fulfilment. While destinations may vary and paths may change, it is the journey itself that enlightens. Every step towards harmony and joy enriches your soul and aligns you with the Universe's limitless possibilities. Nourish your roots, reach towards the sunlight, and allow the Divine Energy of the Universe to guide and unfold your life.

As you progress through this book, you will encounter insights to deepen your connection with your inner self and the Universe. This book serves as a guide and companion on your journey, illuminating your path towards a more harmonious and joyful life. Embrace the wisdom within these pages and let it light your way.

And because I embrace the present moment, with serenity and appreciation, all my desires are realised, under grace and in perfect ways.

Concepts and Definitions

> *"Energy flows where attention goes."*
> – JAMES REDFIELD

The Law of Vibration: Energy of Motion

THE HERMETIC TEACHINGS, ORIGINATING FROM THE ANCIENT wisdom attributed to Hermes Trismegistus, offer a philosophical framework that explores the nature of the Universe, the mind, and the connection between the material and spiritual realms. One of the key principles in these teachings is the Law of Vibration, which states that everything in the Universe is in constant motion; everything vibrates at a certain frequency. Subatomic particles are perpetually moving, creating vibrations. Your thoughts and emotions generate specific vibrational frequencies that interact with the Universe, influencing the reality you experience. By understanding and applying this principle, you can consciously elevate your vibrational frequency, shaping the reality you encounter and aligning yourself with the energies of the Universe to bring your desires into being.

The Law of Attraction: Shaping Reality with Thought and Emotion

The Law of Attraction suggests that by aligning your thoughts and emotions with what you desire, you elevate your vibrational frequency to match the frequency of your desired outcome. This alignment enables you to transform your reality by drawing experiences and opportunities into your life that resonate with your intentions. Essentially, the vibrations of your thoughts and emotions bring about outcomes.

Manifestation: Making the Invisible to the Visible

Manifesting is the process of consciously raising the vibrations of your thoughts and emotions to impress upon the invisible, universal substance, thereby bringing the invisible into the visible physical plane. It involves the practice of setting clear intentions, visualisation, and maintaining a high vibrational frequency. Manifesting is about believing that what you desire is already yours and feeling the outcome as if it is already happening. This alignment of thought and emotion plays a central role, as the Universe responds to the vibrational frequency of your inner state. Manifesting is a harmonious dance between believing and taking inspired actions, where your inner state shapes your outer world. Remember, in the realm of manifestation, believing is seeing.

The Universe: Infinite Intelligence

When I refer to the Universe, I speak of infinite wisdom—an all-encompassing intelligence, knowledge, and understanding that permeates everything, and the energy that connects us all. In Hermetic teachings, this is known as The ALL—everything that encompasses and binds us. This expression captures a belief that is fundamentally spiritual and not tied to any particular religion. The term "Universe" can be interchangeable with other names such as The ALL, God, Allah, or any other name that resonates with your spiritual beliefs. It is a concept that transcends individual belief systems, inviting a universal connection to the Divine and offering deeper understanding to those who align themselves with this boundless wisdom.

The Chakra System: Wheels of Energy

Chakras, derived from the Sanskrit word for "wheels," are internal subtle energy centres that regulate various aspects of our physical, emotional, and spiritual well-being. These spinning wheels of energy influence different facets of our lives—from physical health and emotional balance to spiritual growth. When these chakras are balanced, they facilitate a smooth flow of energy throughout our body. However, if they become blocked, they can cause physical, emotional, or spiritual dis-ease. Understanding and balancing these energy centres can lead to profound healing and transformation.

Chakra Balance and the Law of Attraction

The chakra system plays an essential role in the process of manifestation and the Law of Attraction. Each chakra corresponds to different facets of our life, and when they are balanced, they create a harmonious flow of energy that enhances our ability to manifest our desires. For example:

The Root Chakra: Resonance of Intention

When balanced, it provides the stability and grounding necessary to feel secure and focus on your intentions.

The Sacral Chakra: Resonance of Empowering Emotions

The centre of emotions, creativity, and passion, which, when in harmony, energises the pursuit of your desires.

The Solar Plexus Chakra: Resonance of Inner Strength

Linked to personal power and confidence, fuelling your aspirations.

The Heart Chakra: Resonance of Healing

Beats to the rhythm of love and compassion, supporting healing and emotional balance.

The Throat Chakra: Resonance of Expression

Your communication centre, enabling you to articulate your intentions clearly to a listening Universe.

The Third Eye Chakra: Resonance of Insight

Key to activating your intuition and seeing beyond the physical realm, helping you visualise your desires.

The Crown Chakra: Resonance of Divine Connection

Connects you to higher consciousness, providing spiritual guidance and a sense of unity with the Universe.

By balancing and activating these chakras, you align your entire being with the vibrations of your desires, thus enhancing the effectiveness of the Law of Attraction. This holistic approach ensures that your physical, emotional, and spiritual states are aligned, making the process of manifesting more effective.

Luella Goethals

Let us embark on this journey together and discover the path to true fulfilment, where the practical and the spiritual blend seamlessly to create a life of inner serenity.

And because I align my thoughts, emotions, and chakras for serenity and appreciation, all my desires are realised, under grace and in perfect ways.

Illuminating Our Path: We Are Beacons of Light

> *"Sometimes our light goes out but is blown into flame by another human being. Each of us owes deepest thanks to those who have rekindled this light."*
> – Albert Schweitzer

HAVE YOU NOTICED THAT CERTAIN MEMORIES ARE MORE VIVID than others, especially over time? These memories are also your teachers.

This profound truth is something I experienced firsthand during my first year of secondary school in England. Sister Mary, with her thick, short hair wavier than usual, probably because she had let it dry naturally, hurried into the classroom. The morning sun streamed through the windows, casting a golden glow on her flannel grey winter robe and flat rubbery shoes. Smiling her warm smile, she instructed us to write down the date, mentioning that the sequence of sevens was special: 7/7/.

Meticulously crossing my sevens French-style, I imagined myself on Boulevard Saint Germain, as we had learned and seen in our French lessons. The aroma of freshly brewed coffee mingled with the scent of warm croissants, while the soft murmur of conversations and the chiming of porcelain cups created a comforting symphony. Sunglasses protected my eyes from the hazy glare, and the waiter's white apron was tied neatly around his waist. Traffic suddenly piled up as a lorry double-parked to unload its boxes. Cars hooted unrestrained. A slender man in blue suit-trousers, a white shirt, and tinted-grey sunglasses stepped out of his car, gesticulating impatiently and muttering to himself.

The unusual spell of silence in Sister Mary's classroom urged me to sit up, shaking me out of my reverie. Sister Mary bent over a record player, her hand poised three centimetres above it, fingers gently gripping the thick white plastic casing around the needle. Thirty pairs of eyes watched the record turning round and round. Relief rippled through the room when Sister finally lowered the needle into position. Music startled the morning stillness. How strange. The lyrics curiously echoed the rhythm in my heart. *I Am a Rock* by Simon & Garfunkel. I found myself drawn into its world. Sister Mary posed an intriguing question, "What would it feel like, or what would be the impact if each of us in this room behaved like cold rocks and solitary islands?"

Eager hands shot up.

"It would be very strange," Susan's voice came from the back of the classroom.

"It is unnatural not talking to anyone," the teacher nodded.

"I think that we would forget many simple things, such as relating with each other," Marie, sitting in the front, answered. She always reserved me a place next to her on the first bench on Fridays, for our lunchtime mass in the gym hall."

Good. And, what about your feelings?"

My eyes wandered to the stream of sunlight pouring onto my school desk, transforming the polished oak into a pool of shimmering gold. I gazed at the glimmer, gripped by my own feelings.

"I would feel sad, even afraid!" replied Julie, sitting to my immediate left.

The nearby voice impelled me to sit up. I looked up and around. Gently moving my desk forward and tucking my chair closer to its edge, I practically squeezed the air out of my stomach, positioning myself so that Simone's broad outline in front of me obscured me from the teacher's view. The discussion was not for me. I needed to remain comfortably invisible, resting on that hard rock of solitude. This was the way I had been raised. My mother set the rules: as Indian immigrants from Kenya, now in London, we kept to ourselves. Blood was thicker than water, and my allegiance to my family came first and alone.

Sighing, I slouched over my desk and began fiddling with strands of my black hair that shone a beautiful auburn in the sunlight. The split ends were numerous and wonderfully translucent. Pinching one split-end with thumb and index finger, I slowly and deliberately peeled it away from its main strand. Another straggled end, another, and then another was ripped apart mercilessly.

As Sister Mary paced the classroom aisle, her eyes swept over her first-form pupils, and her tone bounced off the walls. She reminded me of a protective lighthouse. Facing the class, she became silent.

"This is not the way to be. We are here to interact with each other, lovingly. You are each a beacon of light. Your attitude determines the quality of your light. Shine with caring. Take an interest in each other. Interact lovingly. Shine brightly. Just imagine an assembly of such beacons! Consider this."

As I reflect on Sister's words, I realise that we are here to resonate with happiness. And I truly believe that we are each beacons of light. So, the question is, how do we shine our light?

By understanding and harnessing the Law of Attraction and the power of the chakras, we can align our inner energy with the Universe's vibrations. The Law of Attraction teaches us that like attracts like—our thoughts and emotions shape our reality. When we focus on loving thoughts, we attract the same energy back to us.

The chakras, our body's energy centres, are key to this process. Balanced and aligned chakras ensure that our inner light shines brightly, attracting harmony and fulfilment into our lives. When our chakras are in balance, our energy flows freely, allowing us to resonate in harmony and to manifest our desires.

Let us explore this together, as we illuminate our paths and those of others around us, inside out.

And because I align my inner light with love and compassion, all my desires are realised, under grace and in perfect ways.

The Subconscious Mind

> *"The mind is everything. What you think, you become."*
> — BUDDHA.

BEFORE WE EXPLORE THE ROLE OF EACH CHAKRA AND THE LAW OF Attraction, it is important to understand the power of the subconscious mind. This profound realm within us is where our deepest beliefs and emotions reside, subtly influencing our thoughts, behaviours, and the reality we experience. By understanding and working with our subconscious mind, we can consciously align our thoughts, emotions, and energy centres to manifest our highest aspirations.

We are born in the likeness of the Creator, endowed with the ability to discern and the power to create. We discern with the gift of free will and create with the vibrations of thought and emotion. These elements, combined with the creative powers of the subconscious mind, govern the song and the lustre of our Soul.

The subconscious mind is a mysterious intelligence quietly working in the background, recording, filming, and storing in its infinite database all that you hear, see, smell, sense, taste, and all that you are told, taught, or instructed. It pours information into its bottomless pool with indifference, without bias. Within its depths, it forms patterns and associations. These patterns emerge because thoughts and emotions create threads that link similar experiences and emotions, creating a web of interconnected ideas and memories.

Dominant and similar chains of thoughts bubble to its surface. These can be subconscious links, unsummoned and yet triggered to arise to the top – like the associations in a child's subconscious mind aroused by the refrains of "stupid" or "selfish" or "sulky." The surface can also contain conscious thoughts, summoned by you because you use discernment – like the adult choosing to reflect upon her skills and using her free will to refocus on harmonious beliefs.

This prevailing film of conscious or unconscious thoughts and emotions diffuses into the ethers of the Universe and thereafter, through the mystery of the Universe, takes substance and form in our physical plane. People, places, situations, or things play out on Earth the film upon this heavenly well. These are seen by you as circumstances and synchronicities – the child encounters a string of failures, but in adulthood, as she chooses to focus on her skills, she meets accomplishments. The dominant conscious, unconscious, subconscious, collective, tribal, or inseminated thought is the cause: the circumstance or synchronicity is the effect.

This is also the principle of the Law of Attraction, whereby like attracts like. Reflect upon success, and anticipate bubbles of celebrations. Dwell on failure, and accept films of failure. Beckon thoughts of sadness and notice sorrowful currents. Believe in punishment and situations feed your conviction.

This Law echoes the old adage: you reap what you sow in your subconscious mind. Spiritual teachers often describe this as a garden that fertilises the prevailing seeds of thought—whether they are conscious, unconscious, collective, tribal, or externally planted.

Your purpose in life is to adorn the sanctuary of your subconscious mind with thoughts and desires sensibly selected by you. Use your gift of free will to clear your subconscious well: refocus on harmonious beliefs and fill your creative waters with your ideals.

All your experiences – the anticipated, the unexpected, the good, the bad, and the ugly – are creations of some thought contributing to your growth and evolution. How you deal with circumstances determines the song and lustre of your Soul. Your anger, grudges, and hate resonate into the deepness of our very loving Universe, which does not judge but breathes easily and radiantly with the lighter resonances of compassion, forgiveness, or love. In the grand scheme of the Universe, the human labels of right and wrong, good and evil, are constructs we create to make sense of our experiences. They just are, serving as opportunities for growth and deeper understanding, even in the midst of great challenges and suffering.

The subconscious mind, therefore, plays a foundational role in aligning our chakras and effectively using the Law of Attraction. By consciously filling our subconscious with thoughts that resonate with our highest aspirations, we can better align our energy centres and attract the experiences that nurture our unique spiritual and physical well-being.

Do you prefer to contribute to the absence of light or to enhance the brighter resonances?

And because I am the captain of my subconscious mind, I glow on a cellular level, inside out, under grace and in perfect ways.

True Story; The Stranger

> *"Man, alone, has the power to transform his thoughts into physical reality; man, alone, can dream and make his dreams come true."*
> NAPOLEON HILL

This true story illustrates how the vibrations of our conscious and subconscious thoughts and emotions manifest in our reality, often in surprising ways.

Standing on the edge of the pedestrian crossing in a local town on the outskirts of Paris, waiting for the high-pitched bleeps to sound, Helen rattled her keys in the deep pockets of her long, grey woollen coat. She absently glanced back. The market stall gleamed full with firm yellow and red tomatoes, indigo aubergines, deep-emerald broccoli, sandy carrots, and crisp lettuces. Patrice selected the larger butternut for his client. Another old lady, dragging her empty trolley-bag, joined the growing queue.

The bleeps startled Helen into awareness. She turned around and walked on, slowly, gathering her thoughts for the imminent task. But a bellowing voice behind, growing in volume, disturbed her concentration. Frowning, she absent-mindedly fumbled her pockets for her keys but intuitively turned around to see an unshaven man in a dark trench coat beckoning her. She met his eyes.

"I want to know your name! I want to know exactly who you are!" the man shouted.

She shook off the stranger's words. She quickened her steps and headed towards her bank, just ahead.

The only free cubicle in the bank invited her to sit down. Elbows on the desk, chin resting on praying hands, she closed her eyes and inhaled deeply. Composed, she rummaged around in her bag for her chequebook, her identity card, her latest electricity bill, and the official forms, arranging them neatly on the table. The surname on the papers stared at her. She matched their steady gaze. Black ball pen in hand, she began filling out the necessary documents to erase that marital name.

A musty odour and the sudden movement of air around her space broke the beads of concentration. A man's face peered eerily over her right shoulder – the stranger had followed her into the bank!

"I'll eventually find out who you are!"

Helen held her breath. The bank staff swiftly approached the man and escorted him out of the building.

She drank in those potent words and trembled for many hours.

Helen had recently moved into her own apartment. During the past week, her identity card, credit card, chequebook, visit cards, and bills sprawled the dining table. She needed to see her maiden name on those papers. But more, she ached to know herself, to be her true self, and to find her place in life after twenty years of being in the shadow of a gentle but forceful husband.

The stranger had mysteriously echoed her poignant thoughts, illustrating the powerful vibrations of her conscious and subconscious thoughts and emotions.

And because I align my conscious and subconscious thoughts with clarity, I advance in this life journey, under grace and in perfect ways.

The Root Chakra: Resonance of Intention

"Just as a tree firmly rooted in the earth can withstand the storm, so too can a person grounded in their being endure life's challenges."
– Unknown

THE ROOT CHAKRA, OR MULADHARA IN SANSKRIT, IS THE foundation of your energy system, located at the base of the spinal cord. It governs your sense of safety and stability. When balanced, it radiates a vibrant red, symbolising strength in both body and mind. However, when imbalanced, the root chakra can take on a dull hue and feel "unmoored," leading to insecurity and a tendency to dwell on past painful experiences. This disruption can hinder your ability to effectively engage with the Law of Attraction, which operates on the principle that like attracts like.

To restore balance, release the past and anchor yourself in the present. True alignment, essential for working with the Law of Attraction, can only occur here and now. Embracing the present heals the root chakra, creating a stable foundation from which desired energies can flow.

As you ground your intentions and nurture them with care, a ripple effect begins to manifest in your life, reflecting the growth and stability you have cultivated.

Nurturing Your Growth with Purpose

Visualise yourself as a sturdy tree, deeply rooted in the soil. Just as a tree draws strength and stability from its deep roots, you too can cultivate an inner ecosystem. In this personal ecosystem, you are the architect of what nourishes you and contributes to your growth. Make space in your mind to fully embrace your feelings and thoughts in the present moment. Observe. Be still. Even a few moments of mindful presence can anchor you, keeping you grounded and resilient against external pressures. Clear away both mental and physical clutter to allow your inner ecosystem to flourish. The time is now, and the place is here.

Grounding ourselves enables us to navigate the complex web of societal expectations while staying true to our authentic selves. The root chakra plays a significant role in helping us understand and navigate these societal norms, but it is essential to do so without losing sight of who we truly are. When conflicts arise between societal expectations and our inner truth, the root chakra can become unsettled. Balance and a sense of well-being are restored when we learn to engage with these expectations thoughtfully, without compromising our true selves. By embracing the rules of various communities—whether in work, sports, or other associations—with an open mind, we can remain genuine and true within these frameworks.

In the various communities we inhabit, we take on different roles—whether as a mother, a wife, a woman, or a professional. Each role brings its own set of self-expectations for growth and development. To remain true to yourself, it is important to understand and embrace these self-expectations while staying authentic. As you journey through life, you make decisions with the best knowledge available at each moment. Though hindsight may reveal past mistakes and shortcomings, it is through intentional growth and self-expansion that you evolve into a better version of yourself each day. Embrace the intention to learn and grow, even when faced with criticism. Balancing these roles requires managing them in a way that no single role overwhelms the others and recognising that your well-being is fundamental for fulfilling each role effectively. Just as airlines instruct us to "put on your own mask before helping others," it is important to focus on your own growth and well-being first. Allow others to be inspired by your journey—or not—as you continue to nurture your own path.

Anchoring Your Intentions

Anchoring is the practice of grounding yourself in the present moment, establishing a stable and secure foundation from which to set and pursue your intentions. This practice supports the root chakra, ensuring that your energy **is** centred and balanced. When you are anchored, your intentions become more focused and aligned with your true self, which is fundamental for effectively working with the Law of Attraction.

Start each day with a ritual that connects you to this grounding energy. A spiritual breakfast, as I call it, sets the tone for serenity and alignment. During this time, plant seed-thoughts—intentions of what you desire—in the fertile soil of your mind. These intentions should be clear, precise, and formulated in the present tense, as the Universe exists in the now. For example, instead of thinking, "I want to be happy," affirm "I am happy," or "I am creating a beautiful day." This daily practice nurtures your intentions, helping them take root and grow, thus activating the Law of Attraction to draw those outcomes into your life.

The Power of the "I AM"

As you ground yourself in the present moment, the power of 'I AM' statements becomes a natural extension of this practice, resonating with the Law of Attraction.

The "I AM" statement is a powerful tool in this process and works in harmony with the Law of Attraction. It acts as a tuning fork, aligning your body, mind, and spirit with the vibrations of your desires. When you declare "I AM," you are not just speaking words; you are setting a resonance that influences your reality and attracts corresponding experiences. Therefore, it is important to choose your "I AM" statements wisely. Avoid phrases that diminish your energy, such as "I am tired" or "I am unlucky." Instead, tune your statements to resonate with what you truly want: "I am strong," "I am successful," "I am flourishing."

This alignment through "I AM" statements is further reinforced by writing down your desires daily. Writing is not just an act of documentation; it is an act of creation. When you write down your intentions, you give them form and direction, helping to focus your subconscious mind on what you wish to manifest. The act of writing also clears your mind, making space for new, inspired thoughts to take root, which then energises the Law of Attraction to bring those thoughts into reality.

Protecting Your Intentions

As you set and nurture your intentions, it is important to protect them from external influences. Just as you would protect a young plant from harsh weather, shield your desires from adverse influences and doubt. Keep your goals close to your heart, sharing them only with those who support your vision. This protection allows your intentions to grow in a constructive environment, free from interference, and ensures that the Law of Attraction can operate without obstacles.

Focus on the end result as if it has already been achieved. Feel the relief, the joy, and the inner serenity that come with realising your desires. This focused attention helps to align your energy with the outcome you seek, bringing it closer to manifestation through the Law of Attraction. Remember, the process is not about controlling how your goals will materialise but about staying aligned with the belief that they will.

Manifestation and the Subconscious Mind

Manifestation is the process through which your thoughts, intentions, and desires are transformed into reality. This process is deeply connected to the Law of Attraction, which operates on the principle that like attracts like—meaning your inner state and intentions attract corresponding experiences into your life. The influence of the subconscious mind on this process cannot be understated; it holds the deep-seated beliefs and emotions that shape our reality.

To effectively manifest your goals, it is essential to align your subconscious mind with your conscious intentions. When both are in harmony, your ability to attract what you desire is greatly enhanced. For example, if you consciously aim to improve your career but hold subconscious beliefs of self-doubt, your manifestations may be impeded. Conversely, when your inner thoughts, emotions, and beliefs are consistently aligned with your goals, the Law of Attraction can work seamlessly, drawing in opportunities, people, and circumstances that support your aspirations.

By nurturing and grounding your intentions, and ensuring that your subconscious supports your conscious desires, you create a powerful synergy that facilitates the manifestation of your goals. This alignment ensures that your energy flows freely, enhancing the effectiveness of the Law of Attraction in bringing your desires to fruition.

The Ripple of Your Intentions

As you ground your intentions and nurture them with care, you will begin to notice a ripple effect in your life. Your progress will naturally inspire those around you, not by force, but by the light of your example. Just as a well-rooted tree stands strong and tall, providing shelter and inspiration to others, so too will your grounded intentions create a positive influence on your surroundings.

Consider the example of Nelson Mandela, who declared from his heart in 1963: "I have cherished the ideal of a democratic and free society in which all persons will live together in harmony and with equal opportunities." Mandela's vision was clear and vibrant. He spoke as if it were already a reality, and in 1994, his ideal was realised.

The Universe rewards persistence and patience. Its notion of time is not linear but is shaped by your current thoughts and emotions. The deeper your belief and feeling, the closer your thoughts move toward physical reality through the Law of Attraction.

Reflect regularly on your intentions, reinforcing them through writing and mindful attention. By cultivating your inner ecosystem—staying grounded, centred, and focused—you create a fertile environment for your desires to take root and grow. This internal alignment allows the Law of Attraction to harmonise with your intentions, drawing the experiences, opportunities, and circumstances that resonate with your true self. In this way, you are not merely reacting to the external world, but actively shaping your reality from within.

And because I align my intentions with purpose and direction, my desires manifest under grace and in perfect ways.

True Story: Uncle's Energy

> *"The greatest weapon against stress is our ability to choose one thought over another."*
> – WILLIAM JAMES

Sometimes, revisiting familiar places brings unexpected reflections and insights. During a visit with my uncle, what began as a simple encounter opened up deeper layers of memory and emotion. As we navigated this moment together, I found myself drawn into a journey that blended the past with the present, revealing quiet lessons about the strength in healing and the importance of embracing life's uncertainties.

The doorbell was broken. I turned the brass knob and entered the house, as Uncle had instructed over the phone. The house was eerily quiet. In my mind, I heard the sounds of his young children laughing and running around. But these were images and sounds as I had remembered them when I was fourteen, more than thirty years ago.

Taking a couple of steps forward and seeing that the door on my immediate left was wide open, I walked into the room, tiptoeing to avoid the echo of clicking heels on the wooden boards. Making my way to the end of the room, automatically towards the daylight pouring in through the white curtained windows, I settled into a large armchair. Is this the same pink armchair from thirty years ago? It still had its lovely pink sheen, though. The television to my right, a standard cathode ray tube set, definitely used to be in the adjacent room. Is that wooden table one that I should recognise and cannot recall? When did he install that shelf along the length of the wall? Determined to stay absolutely focused on the present moment, I started to examine my fingernails, giving them a mental manicure.

I suddenly noticed Uncle moving in through the door. His shoulders swooped as he practically shuffled his way towards the couch near the entrance, to my right. He sat down slowly and with precaution. With a weary gesture, he wiped away the solitary tear rolling down his cheek.

"I don't want to die. I'm not ready. There are so many things that I want to do," he spoke softly, and the prominent crackle, more distinct now than over the telephone, suggested surgery around the trachea.

I had already met Uncle in my dreams. I was prepared for his degraded physique and his oozing distress. He was the one who instinctively spoke out directly about his concerns, brushing aside all formalities.

"What do the doctors say?" I asked, breaking the long silence.

"I'm making very good progress and I'm responding well to the treatment," he paused. "But I've lost my appetite. It's as if the treatment has burnt my tongue, perhaps even the taste buds."

I nodded, knowing that his flavour for life was barely present. Having meditated earlier on Uncle's chakra energy, his body cells were fast repairing but were vibrantly appealing for him to focus valiantly on his living and healed cells.

"But do you really need to continue the treatment? I sense that you are practically healed. Talk to the doctors concerning your current treatment and even the dosages."

Uncle seemed a little surprised. We had not spoken about his treatment over the telephone and I suddenly realised that I was exposing my intuitiveness, strongly cautioned against by our Catholic upbringing, and Uncle was family.

"Focus only on your good progress," I felt the need to speak slowly and to articulate my words carefully. "But there is something else. Good Friday is approaching and I sense that your emotions are heightened on death." Uncle looked up, a little alarmed at his own transparency.

"It's a beautiful day but it is very cold. Wrap up warm and go into your garden. Someone took great effort to plant all those beautiful daffodils. Take a look around and observe spring unfolding. It is Easter. This is your rebirth. That's all that matters, right now."

"Yes, but what about all the bad news in the world? I put on the TV and it's just so, well, so depressing." He paused and opened up further. "I may even lose my sight - with all the treatment."

"It's time to switch off the news." I hesitated, wondering if he would adhere to centring completely on his one guarded but innermost desire. "There is one more thing. Do not bring attention to illness. Only talk health and healing and see the wonders in the people assisting you."

"Yes. It's a great team. But it takes enormous effort dealing with these people and their blunders. The secretary forgot to schedule the ambulance. The nurse just stabs in those injections!"

"It's time to switch off the news." I hesitated, wondering if he would adhere to centring completely on his one guarded but innermost desire. "There is one more thing. Do not bring attention to illness. Only talk health and healing and see the wonders in the people assisting you."

"Yes. It's a great team. But it takes enormous effort dealing with these people and their blunders. The secretary forgot to schedule the ambulance. The nurse just stabs in those injections!"

As we talked, I shared my insights on dealing with my Godfather's perceived hindrances. Those 'blunders' gained significance through the meaning he attached to them. In teaching Uncle the art of letting go of resentments or issues that did not contribute constructively to his prime objective of healing and living, I reflected on our shared family traits. Bitterness and holding grudges were ingrained in our family patterns. Uncle and I, entangled in the same Indian-Goan and Catholic culture, were both learning to release these emotions. Surprisingly, Uncle acknowledged and appreciated my wisdom and spiritual approach, giving me the strength to continue on my intuitive path.

He called a year later after that discussion to say that he is 'clear;' the code for perfect health.

And because I ground myself in the present moment and nurture my intentions, I manifest stability and vitality in my life, under grace and in perfect ways.

The Sacral Chakra: Resonance of Empowering Emotions

"You have to feel the pain to heal it."
– Unknown

THE SACRAL CHAKRA, KNOWN IN SANSKRIT AS SVADHISTHANA, is the seat of our emotions, passion, creativity, and sensuality. Situated just above the root chakra, it radiates a vibrant orange hue when in a healthy state, serving as the wellspring of your emotional and creative life. This chakra is associated with the element of water, symbolising flow and adaptability. When emotions become trapped or suppressed, the sacral chakra can darken, leading to digestive issues and emotional turbulence. This imbalance disrupts the free flow of energy, affecting your interactions with the world and your inner sense of harmony.

The sacral chakra is integral to the Law of Attraction, as your emotions are powerful energy currents that influence the vibrations you send out into the Universe. When this chakra is balanced, emotions flow freely and harmoniously, amplifying your creative energy and attracting experiences aligned with your desires. However, when this chakra is blocked or out of balance, the energy becomes stagnant, and you may find yourself attracting situations that reflect this inner disharmony.

To fully appreciate the power of the sacral chakra, it's essential to explore the intricate relationship between thoughts and emotions, as these forces shape the energy you project and, ultimately, the reality you create.

The Energy of Emotion

Behind every thought lies an emotion. Emotions power thoughts, and thoughts, in turn, fuel emotions. For example, when you feel tired, it's not merely a physical sensation; it is driven by a thought, perhaps from a restless night or a challenging day. Recognising this connection is key to understanding how your emotional energy influences the sacral chakra.

Emotions are a form of energy. The word "emotion" comes from the Latin "emovere": *e*, meaning "out," and *movere*, meaning "to move," giving the image of emotions as energy in motion, moving outward and rippling into the environment. This energy is palpable in crowds, such as the joy at a wedding, the deep sadness at a funeral, or the tension during moments of conflict.

Consider your emotions as expressive energy that cannot be destroyed or suppressed. When denser emotions like sadness, pain, grief, anger, anxiety, jealousy, or unforgiveness are left unexpressed or unattended for long periods, they become trapped energies. Without a proper outlet, these energies disrupt your inner harmony, leading to imbalances in the sacral chakra and eventually manifesting as physical or emotional dis-ease. It is often said that the body expresses what the mind represses.

Owning Your Emotions

Certain cultures and generations teach the art of self-control, often denying the expression or even the existence of emotions such as anger, sadness, regret, fear or even passion. However, all emotions are real, valid, and require your attention. Recognise and name your emotions; acknowledge when feelings like anger, sadness, frustration, or impatience arise.

You are not defined by your emotions. Instead of seeing yourself as an angry or sad person, recognise that these emotions are temporary states that you can acknowledge and manage. By owning your emotions, you maintain balance within the sacral chakra, allowing for emotional freedom and personal empowerment. For instance, you might say, "I am angry," "I am feeling sad," or "I am running out of patience." This approach allows you to recognise and take responsibility for your emotions without letting them define who you are. In this context, the "I am" statement is not an intention but a way of acknowledging and taking ownership of your current emotional state.

No one else causes your emotions. People, situations, and circumstances may act as triggers, but they do not control your emotional state. Avoid saying, "You make me angry," as this shifts responsibility away from you.

Instead, clearly express your emotions and the reasons behind them: "I am feeling angry because I perceive that you are not fully contributing to the household tasks." This helps you take ownership of your feelings and communicate the underlying issue, which empowers you to address it directly.

Examine the thoughts that fuel your emotions. For instance, is your jealousy linked to feelings of unworthiness or competition? Is your anger connected to thoughts of powerlessness, rejection, or a lack of affection? When you experience sadness, consider whether you are clinging to thoughts of emptiness or past memories that dampen your spirits. Recognise that you are responsible for your own emotions and thoughts, and no one else has the power to dictate how you feel.

Reflect on how changing your focus can shift your emotional state. By altering your thoughts, you change the emotional charge and begin to heal your sacral chakra.

Restoring Your Emotional Flow

Unpleasant past experiences can deeply embed emotions within the subconscious, creating blockages that hinder the free flow of energy in the sacral chakra. These blockages can disrupt your emotional equilibrium, making it challenging to maintain a constructive and harmonious perspective on life. Staying present and aware of your current emotions, acknowledging that they are real and valid, is key to releasing these emotional blockages.

Engage in practices that help you release these emotions from your system, such as physical activity, journaling, or therapeutic crying. While sharing with friends can be supportive, it is important not to rely solely on them as your outlet for emotional energy. Friends may offer perspectives shaped by their own experiences, which can be helpful for benchmarking and refocusing. However, for a more neutral and supportive approach, consider seeking guidance from a professional, such as a coach, counsellor, therapist, or psychologist, who can help you process your emotions in a healthy way.

As you clear these emotional blockages, your energy begins to flow more freely, aligning you more closely with your true intentions and desires. This unimpeded flow of energy is essential for the Law of Attraction to work effectively, as it enables you to attract experiences and opportunities that resonate with your highest aspirations.

Transforming Your Disempowering Emotions

Anger is often an old, deeply rooted emotion that has been unaddressed over time. By recognising and transforming these emotions, you restore balance to the sacral chakra, allowing for a renewed flow of energy that supports your well-being. This means it is not the person or situation itself that creates your anger but rather how it touches on unresolved emotions from the past.

For instance, you might feel frustrated by a lack of time for personal projects or struggle with aspects of yourself that you know need to change. This frustration can be a superficial expression of deeper, unresolved anger. For example, if you experience ongoing irritation when you are unable to focus on your hobbies, this could be linked to old, unmet needs or past frustrations that have not been fully addressed. Recognising these connections helps in understanding that your current feelings might be rooted in older, unresolved issues.

Recognise that anger, as an old emotion, can manifest in various forms, from minor irritations to deep frustrations, often linked to feelings of lack or unmet needs. This deeply rooted emotion may have origins in past experiences where your needs were not met, leading to a lingering sense of dissatisfaction that persists into the present. Rather than attributing your anger to others, take a moment to identify the true source within yourself. For example, instead of saying, "You are annoying me," acknowledge the internal trigger by saying, "I am annoyed by..."

Understanding that your anger is an old, unresolved emotion—whether it stems from unmet personal goals or a deeper sense of unworthiness—empowers you to take control of your emotions and your life. This approach not only strengthens your mind but also aligns your energy, allowing you to actively participate in your own evolution.

Remember, the Law of Vibration, much like gravity, draws to you people, places, and circumstances that resonate with your internal state. By addressing and transforming disempowering emotions, especially those as deeply rooted as anger, you align yourself with the energies that support your highest good, attracting experiences that reflect this alignment.

The Power of Polarity

The Principle of Polarity, as taught in The Kybalion—a Hermetic philosophy of Ancient Egypt and Greece—explains how mental states can be transmuted into their opposites through trained will. This principle is especially powerful in sacral chakra work, where transforming disempowering emotions into empowering ones can restore harmony and flow. Just as heat and cold are two extremes on the same spectrum, so too are emotions like sadness and joy, or anger and compassion. These polarities exist as degrees of the same essence, differing only in their intensity.

By applying this principle, you can consciously transform a disempowering emotion into its empowering counterpart. For instance, sadness can be shifted to joy by dipping into the treasure chest of happy memories, and jealousy can be transmuted into rejoicing or acceptance. Possessiveness often masks deep tenderness, while grudges, which grasp the heart, can be liberated through thoughts of forgiveness or release. By peeling away the layers of anger, you may reveal compassion. Tones of compassion include gentleness, kindness, and consideration—not necessarily towards others, but first and foremost towards yourself.

Consciously dwelling on high-resonating emotions such as joy, compassion, acceptance, tenderness, or forgiveness elevates your energy and recharges your chakra system. These empowering states are endowing and liberating, creating a flow that the Universe responds to. According to the Law of Attraction, your emanating thoughts and emotions are intercepted by the Universe, which matches and returns them in the form of people, places, or circumstances. Thus, the Universe, through its mysterious workings, reproduces your feelings without bias.

Emotions cannot be suppressed or destroyed, but they can be transformed or transmuted. For instance, just as hot and cold are degrees of temperature, anger and compassion are closely related emotional states. It is important to experience all emotions fully, for in experiencing sadness, joy becomes sweeter. Transform a disengaging sentiment by first acknowledging its presence and then focusing on its empowering polarity. Shift anger into compassion, emptiness into closure, criticism into acceptance, or disbelief into the mindset of "believing then seeing."

Recognise that while emotions are natural, they can be guided towards their more constructive and healing forms.

Potential Polarities

The following are examples of transmuting powerless states to their potential empowering resonances:

    Addiction to Balance
    Agitation to Harmony or Peacefulness
    Anger to Compassion
    Betrayal to Release
    Bitterness to Sweetness
    Blame to Ownership
    Competition to Cooperation
    Confusion to Clarity
    Criticism to Appreciation
    Dependency to Balance
    Disbelief to Belief
    Emptiness to Fulfilment
    Fatigue to Vigour
    Fear to Courage
    Grudges to Forgiveness or Release
    Greed to Generosity
    Harassment to Peacefulness
    Hatred or Resentment to Love
    Ignorance to Understanding
    Illness to Healing
    Isolation to Connection
    Jealousy to Contentment or Appreciation
    Lies to Truth
    Manipulation to Openness or Integrity
    Mistakes to Lessons
    Pain to Soothing or Comfort
    Pity to Compassion
    Possessiveness to Tenderness
    Problems to Solutions

Sacrifice to Desire or Willingness
Selfishness to Caring
Shame to Self-Respect
Suffering to Relief
Worry to Reassurance

Addressing Your Intense Emotions

Your intense, disempowering emotions attached to another person or a situation require attention and transformation. It is important to disengage your thoughts from the person or circumstance. The power to change outcomes and manifest your desires comes from managing your thoughts and perceptions. While you cannot impose your ideals on others, you can inspire them by embodying noble principles such as success, accomplishment, honesty, unconditional love, compassion, understanding, forgiveness, truth, and tenderness.

By prioritising your own well-being—such as happiness, serenity, or closure—you shape your destiny and move forward harmoniously. As you evolve, embrace the lessons learned from experiences of deep hurt and sadness, as these bring valuable insights for closure. Failure, too, provides experience and knowledge.

The key lesson is to avoid becoming a victim of others' choices or behaviours. People act and make decisions according to their free will, just as you must use your own free will to move forward, releasing hurt and pain in exchange for serenity. Others will either align with your serenity or follow their own paths.

The following exercise is a proven and powerful method to release pain attached to a person or a past situation. While you cannot change another person, you can release your own distressing emotions to evolve joyously and with inner peace. When performing this exercise, refrain from all judgment. In the absolute, there is no right or wrong, nor any concept of wrongdoing, as all simply is. This exercise is not about condoning or excusing harmful actions but about freeing yourself from the burden of unresolved emotions to find inner peace.

Recall a past experience that still evokes strong, disempowering emotions—perhaps feelings of manipulation, exhaustion, helplessness, betrayal, or being outwitted. Imagine this scene anew, but this time, infuse it with the qualities you aspire to embody: compassion, tenderness, humility, or understanding. In this revised mental scene, you might choose to remain composed and observant, or assertively express your perspective with grace.

Visualise this new version vividly and repeatedly, allowing it to resonate deeply within you. As you replay this transformed scenario, ensure it leaves you with a sense of peacefulness and fulfilment. Let this renewed scene fully permeate your mind, body, and soul, reshaping how you perceive and respond to similar situations in the future.

The past, present, and future are interconnected and exist simultaneously in terms of the Laws of the Universe. When you focus on your desired outcomes in the present, you send ripples through both the past and the future. While you cannot change the past, you can refresh its influence on your present and future. By consciously directing your thoughts and intentions now, you reshape how past experiences impact you, allowing you to create a reality that aligns more closely with your true desires. This ongoing process not only aligns your actions with your goals but also enhances the manifestation of your intentions, ensuring that your past, present, and future selves work harmoniously to create the reality you envision.

Through this meditation, you are also creating harmonious spiritual rapport with the person(s) involved. You are not focusing on absolute right and wrong, but rather rehearsing in your mind the display of your finer qualities, behaviours, and principles. You have learned the lessons and are evolving for the benefit of your higher or spiritual self. Like a path forged consistently and steadily in a forest, your mind responds accordingly when you next encounter the person(s) or when faced with similar challenging situations.

Through this exercise, you are learning not to be the victim of people or circumstances. Instead, you are learning to find a solution by focusing and centring on yourself. By taking care of your inner ecosystem and establishing mental boundaries, you are shifting reality and activating the Law of Attraction, which is highly sensitive to emotions and thoughts. Remember, like attracts like. Transmute your emotions to empowering frequencies, and witness miracles.

Always bestow thoughts of compassion upon loud, opinionated, judgmental, or critical persons. For the justice of their deeds is served and returned, magnified to them, the source, by the Laws of the Universe.

There are many truths and perceptions, and thus, there is no absolute right or wrong path. All experiences offer enlightenment. People often serve as guides, reminding you of the qualities, values, or principles that you pursue or avoid along your path of evolution, as a traveller in life. For example, if you abhor but consistently encounter criticism, the message is for you to focus on demonstrating appreciation. Gossip, the unconstrained conversation about another person, signals for you to speak of others in their presence and with their participation. Dishonesty announces that trustworthiness requires your immediate demonstration and consideration. Be the change you desire. Dwell on resonances of happiness or serenity. Place attention on people's strengths and qualities, and notice these thoughts reciprocated by Divine Law.

Transforming Jealousy

If you are feeling jealous, acknowledge this emotion by writing it down and then find its opposite. A potential opposite is acceptance. Begin by accepting where you are right now and what you have. Grant others the authority to be where they are, who they are, and what they have. Focus on your own feelings, as the Law of Vibration responds to your vibrations alone. Your soul cannot see the other person but only what you are feeling. It cannot discern the reasons behind those feelings. Therefore, start by appreciating yourself and what you already have. Take note of the small things around you that bring you a sense of contentment and peace. Allow the Law of Attraction to work through your subconscious by tending to your inner garden with care and appreciation.

Transforming Betrayal

Feeling betrayed is deeply painful because it often represents a misalignment with your values or expectations. Perhaps you reorganised your time to be with someone who was eventually not available, leaving you feeling undervalued or unimportant. However, it is important to recognise that you cannot change another person to make yourself feel better, more relaxed, or more aligned. Your soul does not focus on the actions of others; it only registers your feelings. Your dis-empowering emotions, when unaddressed, accumulate in your second chakra, creating blockages that affect your vibration and overall well-being.

To manifest growth and maintain vibrational clarity, it is essential that your emotions remain clear and unclouded. If you feel hurt by another person's actions or inactions, recognise that this drains your energy because you are inadvertently plugging into their energy rather than tending to your own. By doing so, you weaken your own vibrational power and hinder your ability to align with your true desires.

The pain of betrayal is profound, but it can be transformed by making choices that resonate with your true desires and priorities. Reflect on why this betrayal occurred and identify what was missing in the relationship. This reflection is beneficial because it helps you understand the underlying issues and what steps you can take to realign with your vibrational priorities. Often, the vibrations in the relationship were mismatched, so betrayal becomes less about what was done to you and more about finding your own alignment. Focus on tending to your inner ecosystem and using your free will to move forward. This involves prioritising your growth and transitioning to a vibrational frequency that brings you peace, rather than casting blame or throwing stones at others for actions that were not a vibrational match for you.

By taking these steps, you reclaim your power and refocus your energy on what truly matters to you. This transformation is not just about overcoming the pain of betrayal but about becoming a vibrational powerhouse, fully aligned with your true self and the path of your life journey. Use your free will to move forward, ensuring that your vibrations are in harmony with your desires and your highest good.

Embracing Your Passions

Your passions are the activities that genuinely uplift and invigorate you, naturally enhancing your vibrational frequency. These pursuits, whether they involve developing your skills, engaging in hobbies, or immersing yourself in activities you deeply enjoy, play a significant role in your personal growth. Whether it is sports, music, dance, cooking, entertaining, and artistic expressions like painting, photography, and writing are all examples of how you can connect with your true self. By regularly incorporating these passions into your life, you align more closely with your higher self and resonate with the higher frequencies that link you with the energy of the Universe.

Nurturing Your Creativity

Creativity is an extension of the sacral chakra, a vibrant energy centre that enables you to resonate at extraordinary frequencies. This energy centre is not confined to traditional artistic pursuits; rather, creativity can be expressed through any activity that brings you joy and a sense of fulfilment. Whether you are designing a garden, mowing the lawn, preparing a meal with care, finding beauty in everyday moments, sailing, or walking, these acts of creation draw upon the profound energy of the sacral chakra.

Even if you do not view yourself as inherently creative, the sacral chakra encourages you to explore and experiment with various forms of self-expression. Engaging in activities such as those mentioned permits you to stay in the present moment and then perceive the world with renewed appreciation. This perspective transforms ordinary moments into opportunities for growth and creation.

By acknowledging and transforming disempowering emotions—such as anger or jealousy—into their empowering counterparts like compassion or appreciation, you align more closely with your true desires. This alignment not only restores balance to the sacral chakra but also enhances your vibrational frequency, attracting experiences that resonate with your highest aspirations. Engage in practices that release and heal emotional blockages, such as creative activities or mindful reflection, to maintain balance and enhance your vibrational frequency. Nurturing your passions and embracing emotional flow creates a fertile environment for personal growth and manifestation.

And because I flow with empowering emotions, I am aligned, under grace and in perfect ways.

True Story: Martha's Fear

> *"Courage is not the absence of fear, but the triumph over it."*
> – NELSON MANDELA

When an unexpected opportunity arose, Martha found herself wrestling with inner fears and doubts. Despite her determination, these emotions lingered, testing her confidence. This true story delves into how Martha navigated this challenge, ultimately finding a way to turn her fear into a source of growth and strength.

Martha was concerned about her promotion. She forced herself to think constructively, yet fear stubbornly and persistently lingered in the back of her mind, presenting her with reasons for failure. She finally acknowledged this emotion and decided to personify it.

Fear appeared as a shapeless form lurking in the shadows, waiting to trip her up. She could not see it clearly, as it stayed wilfully in the darkness. Though it had never harmed her, its presence felt unhealthy.

She recognised how it drained her father, who often identified potential complications and was conditioned not to travel on the metro alone in the evenings. Strange people, apparently, frequented these passageways at night.

Martha wondered if her fear assisted her in any manner. To her surprise, she realised it urged her into action. The initial ingredients were not obvious, but her fear encouraged her to anticipate and project her thoughts and actions into the future.

Imagining how to make this fear smaller and brighter, she suddenly realised that all her arguments for finalising that increase in salary were in her head. Her objectives were quite distinct.

Martha had inherited the conviction to be watchful in life. For her, fear was something dark lurking in the shadows, ready to surprise her unawares. In personifying fear, she realised that this principle belonged to her father. By courageously acknowledging her fear directly, rather than cautiously adjusting to its existence like her father, her arguments came audaciously out of the shadows and into the light.

And because I face my fears and focus on light, I move forward in life with confidence, under grace and in perfect ways.

The Solar Plexus Chakra: Resonance of Inner Strength

"And the time came when the risk to remain tight in a bud was more painful than the risk it took to blossom."
– Anaïs Nin

The solar plexus chakra, known in Sanskrit as Manipura, is the centre of personal power, confidence, and self-esteem. Positioned above the sacral chakra and below the navel, it radiates with the vibrant energy of the sun, embodying transformation and dynamism. When in need of healing, its glow may appear tarnished, signalling the necessity to release past burdens, reclaim your self-worth, and fully embrace your true potential.

When we reclaim our personal power and embrace our true potential, we elevate our vibrational frequency, attracting experiences and opportunities that align with our intentions, thereby activating the Law of Attraction.

Releasing the Past

Releasing the past can be incredibly challenging, yet it is essential for freeing the solar plexus chakra and reclaiming our personal power. By letting go of past burdens, we create space for new experiences and opportunities to enter our lives. Letting go is not always straightforward, but it is essential for our growth and well-being. As we reflect on our experiences, we must acknowledge the pain and emotions that may still linger. Only by facing and processing these emotions can we truly let go and create space for new and wonderful experiences to enter our lives.

There are many methods to release the past; for example, through the exercise of rewriting or visualising the past event as you see in retrospect and how it could have played out given your introspection. Another powerful way is to be very much in the present moment and to tend to your ecosystem, your garden of desires and aspirations.

As you let go of past hurts, you free up energy that can be redirected towards manifesting your current desires. The less you are tethered to what has been, the more powerfully you can focus on what you wish to attract into your life now.

As life is a journey, as you drive on ahead, the past becomes more and more distant and even smaller as you look in the mirror. Your present location is where you are right now, and you cannot be anywhere else apart from being in the present moment. However, you can look ahead and be open to new experiences and learnings. If you stop to look back for too long, you may find yourself in resistance to your flow and begin to overanalyse the past. While you may not have the answers to why a situation occurred, remember that the answers always come with time and movement. It is good to analyse past situations, but analysis can lead to paralysis when all your soul requires is growth and expansion by taking life as a journey. You can create your reality and bend destiny, as we have learned through timeless stories. Sleeping Beauty, for instance, did not die but was placed into a deep slumber for one hundred years. Cinderella maintained her posture of compassion and love, which were eventually matched. The Law of Attraction is patient and matches your consistent and dominant vibrations—always. Meaning, release pain in exchange for peace and serenity.

As we release the past, we must also be kind and gentle with ourselves. Healing is a process, and it may take time to fully let go. But with each step we take, we become lighter, freer, and more open to the beautiful and unexpected gifts that the Universe has in store for us. In doing so, we free the solar plexus, renewing our power and vitality.

Nurturing Your Self-Esteem

Self-esteem reflects how you value and appreciate yourself, independent of external achievements. This inner recognition strengthens your solar plexus chakra, fostering the confidence needed to navigate life with assurance. It is about recognising your inherent worth and knowing that your value is not contingent on external validation. You cannot expect others to constantly nourish your self-esteem. This is your energy centre, and it is your responsibility to recognise your worth. Relying on peers, managers, partners, or friends for validation creates a dependency that diminishes your self-empowerment.

When you appreciate your inherent worth, you are less likely to seek external validation or approval. This internal recognition strengthens your solar plexus chakra and aligns you more fully with the energies that attract extraordinary experiences and relationships into your life.

Your self-esteem is powered by understanding your values, strengths, qualities, and beliefs. It remains strong even when faced with external criticism and judgement. By appreciating your intrinsic value, you reinforce your solar plexus chakra, which aligns you with the Law of Attraction, drawing remarkable experiences and relationships into your life.

Building Your Self-Confidence

Self-confidence is the belief in your abilities and judgment, closely tied to the solar plexus chakra. As you trust in your capacity to succeed, this chakra radiates more brightly, drawing experiences that align with your highest intentions. Confidence means learning from the past and embracing the journey of life. There are no absolute good or bad decisions; each choice is an opportunity for growth and self-discovery.

As you trust in your abilities, your solar plexus chakra radiates more brightly, making you a stronger magnet for the experiences and opportunities that align with your intentions.

The best decisions are made through thoughtful consideration of their potential consequences. By carefully reflecting on possible outcomes, you align your actions with your highest good. This process strengthens your trust in yourself and your ability to navigate life's challenges. Recognising opportunities for growth and expansion, and viewing every experience—whether perceived as success or failure—as a valuable lesson, contributes to your development as a traveller in life. Self-confidence is essential for energising your solar plexus chakra and activating the Law of Attraction, as it aligns your vibration with your desires, making you a magnet for success and uplifting experiences.

Embracing Your Values

Values represent the core aspects of life that resonate most deeply with you, such as peace, success, or spirituality. Living authentically in alignment with these values strengthens your solar plexus chakra, empowering you to attract energies that mirror your inner truth. Your personal list usually includes at least five principal values. Each person has a unique set of values, contributing to the richness and diversity of our world. While some values may seem more universally esteemed, it is important to respect that everyone is on their unique journey. Instead of condemning others and thus diminishing yourself, live true to your values and shine brightly from the inside out, as demonstrated by many of our seers and spiritual masters.

Values and priorities can change over time depending on your desires in life. As you evolve, your experiences shape your priorities and values. This fluidity is natural and reflects your growth and evolution, and the experiences necessary for your soul. This change signifies your alignment with your current aspirations and your current life path. When you live authentically and align your actions with your current dominant values, you emit a powerful vibration from your solar plexus that attracts similar energies into your life.

Belief and the Power of Focus

Your belief system is built on ingrained habits and thoughts. By focusing on empowering beliefs, you can transform your inner reality, strengthening your solar plexus chakra and aligning yourself with outcomes that reflect your highest potential. Changing these habits or beliefs is possible by changing your focus of attention. Beliefs, convictions, principles and truths are all interchangeable terms for thoughts. By consciously choosing empowering beliefs, you can transform your inner and outer reality.

Your beliefs shape your perception of the world and your place within it. They influence your actions, reactions, and overall mindset. By focusing on empowering beliefs, you strengthen your solar plexus, your resolve and align yourself with empowering outcomes. The solar plexus chakra holds your beliefs, reflecting and energising your personal power and confidence. For example, believing in the importance of being present in the moment allows you to experience life more fully and reduces anxiety about the future. Similarly, viewing life as a journey filled with opportunities for growth and learning can shift your perspective from fear of failure to embracing each experience as a valuable lesson.

To nurture a belief system that supports your highest potential, regularly evaluate and adjust your focus. Challenge beliefs that no longer serve you and replace them with empowering ones. Remember, where attention goes, energy flows. For instance, if you hold a limiting belief such as "I am too old to learn," recognise and transform it into "I am always capable of learning and growing." This practice not only enhances your self-esteem but also aligns you more closely with a better version of yourself, enabling you to manifest your desires with greater ease.

Harnessing Your Strengths

Your strengths—whether they are innate talents or developed skills—are your personal power tools. By actively engaging these strengths, you energise your solar plexus chakra, allowing your inner power to shine through in every aspect of your life. Your strengths are the unique skills and talents that you naturally excel in or have developed through dedication and practice. These strengths are your personal power tools, enabling you to navigate life with confidence and self-assurance. By consistently leveraging these abilities, you bolster your self-worth and enhance your overall well-being.

Each strength contributes to different facets of your life. Organisational skills bring order and harmony to your environment, while analytical thinking allows you to approach complex situations with clarity and insight. Artistic talents infuse creativity and beauty into the world, and effective communication helps you articulate your thoughts and ideas with compassion and precision. Concentration focuses your energy, guiding you toward your goals, while sociability fosters connections and nurtures a sense of community. Creativity drives innovation, allowing your imagination to flourish.

These strengths are intimately connected to the solar plexus, the centre of personal power and self-esteem. When you recognise and actively engage your strengths, you energise this chakra, aligning with your inner power and enabling you to take confident, purposeful steps. A balanced solar plexus allows your abilities to shine naturally.

Cultivating Your Qualities

Qualities are the intrinsic attributes that define your character and guide your actions. By cultivating and embodying virtues such as compassion and determination, you fortify your solar plexus chakra, empowering you to face life's challenges with grace and resolve. These inherent traits determine how you interact with the world and can be consciously cultivated by observing and embodying the virtues you admire in others.

For example, determination is the inner drive that propels you toward your goals, even when faced with challenges. Compassion involves deeply understanding and empathising with the feelings of others, enriching relationships and developing deep connections. Patience is the ability to navigate life's challenges with serenity, trusting that the answers you seek will unfold in their own time.

These qualities are also closely tied to the solar plexus chakra. By embracing and nurturing these traits, you strengthen your personal power and your solar plexus, empowering you to approach life's challenges with grace and determination.

Illuminating Your Solar Plexus

Each day, take a moment to reflect on your actions and observe which of your values, beliefs, strengths, or qualities were illuminated. This daily practice nurtures your solar plexus chakra, reinforcing your self-esteem and aligning you more closely with your inner power. For example, ironing your clothes in the morning reveals your organisational skills and appreciation for a well-presented appearance, while a spontaneous visit to friends highlights your joy in meaningful connections.

Remember, others might not always understand the reasons behind your actions, as they see only the surface. By recognising and celebrating these moments yourself, you not only reinforce your self-esteem but also pave the way for a more empowered and fulfilling life.

Embrace each day as a canvas to express and cultivate your most radiant self.

Transforming Shame

Judgement, criticism, and societal expectations can trigger shame, leading us to question our worth. However, by embracing self-acceptance and compassion, you can transform shame into personal power, liberating your solar plexus chakra and enhancing your ability to manifest your desires. Shame often stems from internalising these external pressures, causing us to feel unworthy. To release shame, recognise it as an external imposition that does not define your Divine Essence. Your worth is inherent and Divine. Embrace your unique essence with self-acceptance and compassion, seeing it not as flawed but as a distinct and integral part of your being. By doing so, you transform shame into self-acceptance and self-compassion, liberating your solar plexus chakra and enhancing your personal power.

Shame can block the flow of empowering vibrations and hinder your manifestations. The shame of being divorced, the shame of being overweight, the shame of failing—these are just a few examples. By releasing shame, you align yourself with higher resonances, making it easier to attract the life you desire. Your solar plexus chakra, as your personal beacon of light, shines brighter when you embrace self-acceptance. This transformation boosts your alignment with the Law of Attraction, facilitating the manifestation of your dreams.

Allow your values, beliefs, strengths, and qualities to take their rightful place within your mind, body, and soul. By nurturing these aspects, you not only strengthen your solar plexus chakra but also create a foundation for a life filled with empowerment, fulfilment, and radiant energy. Judgement and criticism are verbal undulations that penetrate the heart and contribute to shame. The intention behind these statements—whether to hurt or to demonstrate power and control, consciously or unconsciously—gives these words their power. They are painful and uprooting as they draw attention to a darker side of you that may or may not exist. For example: you are selfish, you are stupid, you never listen, you are negative, you are manipulative, you are tiring, you are slow, you are clumsy, you are disrespectful, you pump my energy, you are intrusive, or you harass others. These are phrases that we may have all heard at some time. Their regular replay in your mind creates disturbances in your subconscious garden.

Ignoring the pain strengthens shame and brings internal dis-ease. This internal struggle dampens the solar plexus and your self-worth. Moreover, marinating in painful waters can tempt you into a prolonged attitude of feeling "victimised." Playing the role of the victim can eventually contribute to an unconscious self-image of helplessness. Clear actions or decisions are unlikely when immersed, consciously or unconsciously, by these troubled waters. The words were dealt out for a reason and require your careful consideration. Surmount ashore, and with calm detachment, consciously consider these waters as a powerful lesson for your personal growth and transformation.

You ultimately have the choice to reject the feedback. You can also mentally return or throw back the words to the source. This does not hurt the person in any way but is a method of returning unwanted and unsolicited vibrations. This action detaches and releases the person from your mind and being. You are not saying that the person is right or wrong. You are merely bringing closure to a situation that no longer serves you, as evolution and serenity become your priority. You are exercising your free will, the ability to discern, your greatest gift in life.

Recognise when you have trespassed or crossed boundaries. Your own grievances might indicate that your thoughts lack clear purpose and direction. This awareness can help you realign your focus and intentions, allowing you to address your concerns constructively and maintain healthy, respectful interactions.

Consider your desires, growth, and expansion. Use your free will to command your soul, and yours alone. But know, in having to make difficult decisions for others who do not possess the ability to discern clearly, you are spiritually guided—especially when you learn the lesson to detach from external judgement.

Break away from the circle of judging and criticising. Think of the evolution of your soul, that deepest part of you, and be the light for others through your intentions and wisdom. In doing so, you nourish your self-esteem and self-acceptance with compassion.

And because I glow brightly from within, my desires take form, under grace and in perfect ways.

Luella Goethals

True Story: Bringing Light to Shadows

"The wound is the place where the Light enters you."
— RUMI

There are moments in our youth that quietly shape the course of our lives, challenging us to confront deep-seated emotions and cultural expectations. This story reflects on a personal chapter from my past, where the struggle for self-acceptance and the weight of silence played significant roles. As you journey with me through these memories, I invite you to consider how our experiences, no matter how challenging, can lead us towards healing and inner growth.

Hawking the movements of the squash ball, the force of my weight smashed it into the far corner of the white stone wall. That felt better. With supple wrists and agile feet, the rebounding ball was always in sight, ready for the strike. Few words were spoken during that impromptu morning lesson on the top floor of the residence hall.

With eyes closed tight, the shower water massaged my aching shoulder muscles. Altering position, the water now ran soothingly down my back. It was easier to breathe and think clearly with the steam. Colleagues had probably left my usual seat in the lecture hall free for me. But this time, there was no lateness; just an intentional absence to reflect before lunch.

The air was still crisp for March. He arrived, wearing his usual long grey duffle coat. As always, he was so relaxed, as if nothing could possibly shatter his world. He saw something in my expression. He joined me on the bench, shoulders hunched. He waited so quietly and patiently for me to speak.

"It's positive."

The hooting of cars far off reached Bloomsbury Square. Footsteps hurried in the street yonder. The legs of a man passed in front and paused not far off as he threw the remains of his sandwich on an already heaped bin.

"Yes, I know," he almost whispered. "The condom broke. I had a suspicion."

He moved so close that I looked up to meet his eyes and words. "I'm not ready to be a father, not at nineteen."

"But it's against our religion and it doesn't make sense. In fact, nothing makes sense."

My gaze wandered to a distant and hazy future.

"Can you see us as parents?" More silence. "We're still growing up."

His question added to the turmoil in my mind. Deeds of sinfulness and feelings of guilt stormed my head. The inevitable hostility charged at me. The odour coming from the egg sandwich on the bin was suddenly nauseating. There was only one safe choice; my private punishment with that God in the heavens, rather than an open and public stoning by parents and family, as is the Indian tradition.

That abortion clinic is part of my past resonances of deep shame and guilt. We were almost fifteen in that room. Our anonymity and silence shrouded the essences of our young and spirited hearts. We were deemed sinners and impure. But I wanted a relationship with myself and a chance to get to know the real me, rather than transfer shame, bitterness, and resentment to a growing child. I needed to stop the pattern.

But the secrecy of that deed slowly, ever so slowly, festered internally. The traits of antagonism and criticism that I abhorred and recognised in my tribe soon after became an integral part of my young adult personality.

That university year came to a close in June. I had failed my final set of exams. It was time to return home. I packed my affairs in silence and slipped away, avoiding all farewells. I was psychologically prepared for the inquisition at home.

As we sat down in the dining room, I listened intently to my parents' questions. I responded with simple yeses or noes, and offered a few shrugs where necessary. Internally, my mind cautioned me to guard my emotions and avoid any reaction. Having navigated this delicate interrogation, I ascended the stairs to my old bedroom, firmly closing the door behind me and locking away the secret in the depths of my heart.

This story, an echo from my past, illuminates the profound struggle between shame and self-acceptance regarding the choices I made for my body and my journey in life. For many years, I kept this secret to myself. At that time, there was no support system available to me, but I am grateful that today, in many parts of Europe, particularly in France where I currently reside and write, support is available to young women facing such challenging decisions that are both emotionally and physically demanding.

Everything happens for a reason, though I did not see it that way back then and simply went through life as it came. Through learning to work with the Laws of the Universe, I have come to understand that excessive analysis can lead to paralysis. Insight often comes with time. In hindsight, my experience was a lesson in learning to trust others, build relationships of trust, and respect the expertise of others.

As you reflect on this narrative, recognise that your journey is uniquely your own. By embracing the full spectrum of your experiences, you lay the foundation for a future enriched by self-acceptance and personal growth. You are also contributing to transforming ancestral patterns, turning shame into resilience.

Infinite wisdom now opens the gates to my Divine healing and evolution immediately, under grace and in perfect ways.

The Heart Chakra: Resonance of Healing

> "The best and most beautiful things in the world cannot be seen or even touched—they must be felt with the heart."
> – Helen Keller

THE HEART CHAKRA, OR ANAHATA IN SANSKRIT, IS THE CENTRE of healing, love, compassion, and connection. Positioned just below the sternum, this emerald-green chakra serves as a bridge between the lower and upper chakras, harmonising emotional and spiritual balance. The term Anahata translates to 'unbroken,' highlighting its essential role in emotional and spiritual healing. Even in moments of deep personal fragmentation, the heart chakra facilitates profound healing at both the emotional and spiritual levels, restoring balance and supporting personal growth. In its optimal state, the heart chakra radiates a vibrant emerald green and moves fluidly. However, when healing is required, it may take on a blue hue, and in the presence of intense emotional distress, it might appear 'wounded.'

When this chakra is balanced, you naturally attract inner peace, resonating from the inside out."

Healing with Serenity

Healing the heart chakra and aligning with the Law of Attraction begins with a journey of deeply connecting to the Universe's loving essence, free from judgement. Manifestation unfolds from a place of inner, where the heart chakra's balance attracts people, situations, and opportunities that resonate with your state of being. When the heart chakra is balanced and flowing freely, the Universe responds to your vibration of healing and serenity. This alignment attracts people, situations, and opportunities that resonate with your current state of being. Manifestation involves bringing your desires into reality by setting clear intentions and feeling as though your desired outcome is already unfolding, all while maintaining a profound sense of inner peace.

Serenity is inherently your right. Relying solely on external validation for inner peace can obstruct your connection with the Universe, which operates beyond judgement. True freedom comes from recognising and tapping into your inner reservoir of wholeness. When you align with the Universe from a place of completeness and serenity, you naturally attract experiences and individuals that reflect this inner state. Inner peace organically leads to serene experiences.

Experiencing pain or distress can impede the flow of the heart chakra and obstruct manifestation. The principle of "like attracts like" suggests that focusing on pain and unmet expectations can hinder alignment with higher vibrational energies. By focusing on healing and working harmoniously with the Universe, you attract experiences that resonate at a higher frequency. All emotions are valid and should be acknowledged. The heart chakra is closely tied to experiences of pain, sorrow, and emotional stress. Prolonged focus on these lower resonances can attract more of the same. Your subconscious mind and the Law of Vibration reflect what you predominantly feel. Lingering on emotional stress can establish it as a dominant pattern in your life.

Pain and distress can impede the flow of the heart chakra, obstructing the manifestation process. Shifting your focus from pain to healing is essential for maintaining alignment with higher vibrational energies, allowing you to attract experiences that resonate with your desired state of being. All emotions are valid and should be acknowledged. The heart chakra is closely tied to experiences of pain, sorrow, and emotional stress. Prolonged focus on these lower resonances can attract more of the same. Your subconscious mind and the Law of Vibration reflect what you predominantly feel. Lingering on emotional stress can establish it as a dominant pattern in your life. For instance, if you are experiencing hurt, the Universe responds solely to your vibrational state, not to the behaviour of others. While grieving the loss of a loved one is natural, clinging to sorrow can hinder both your personal development and the spiritual progression of the departed soul. Instead of fixating on loss or pain, shift your focus toward the resonance of appreciation.

The resonance of appreciation is one of the most powerful forms of healing.

The Resonance of Appreciation

Focusing on appreciation and recognising the blessings in your life aligns you with the grace of the Universe—a harmonious flow of energy that guides you towards extraordinary outcomes and synchronistic events. Appreciation operates at a high vibrational frequency, attracting even more reasons to be grateful and deepening your connection to the Universe's benevolent flow. As you generate this feeling of appreciation from within, you will see it reflected in the world around you. The vibrations you emit ripple outward, and the Universe mirrors this harmonious energy, reinforcing the flow of grace in your life."

As you generate this feeling of appreciation from within, you will see it reflected in the world around you. The vibrations you emit ripple outward, and the Universe mirrors this harmonious energy, reinforcing the flow of grace in your life.

Balancing Giving and Receiving

The heart chakra thrives when there is a delicate balance between giving and receiving. Embracing both aspects ensures that the energy within this chakra flows freely, allowing you to attract harmonious experiences that reflect this balance. While many cultures place great emphasis on giving—whether of time, resources, or support—it is equally important to master the art of receiving. By accepting gifts and kindness graciously, you open yourself to the full spectrum of balance, allowing you to appreciate the small miracles the Universe offers, like a smile or a kind word.

Balance also involves embodying the qualities you wish to receive. For instance, if you seek respect, extend it liberally without expecting it from the same person. What you send out into the world tends to come back to you. While respect is a value upheld across many cultures, when it is demanded, it can shift from a genuine exchange to an obligation, infringing on free will. People's actions are shaped by their experiences and surroundings, influencing their responses. Growth and learning unfold through the cycles of giving and receiving.

Breathing the Vital Force

The heart chakra, associated with the element of air, thrives when the vital force of breath flows freely. Mindful breathing can significantly enhance this flow, releasing stress and inviting calmness. To cultivate this, sit upright with relaxed shoulders and repeat the words: relax, relax, relax. Engage in deep, conscious breaths—inhaling deeply, holding the breath momentarily, and exhaling slowly. Proper posture enhances oxygen flow, which is crucial for maintaining balance and supporting the health of your heart chakra.

As you breathe out, release tension; as you breathe in, invite peace. By aligning with the Law of Vibration, focus on sensations of relaxation and well-being. Regular practice of this simple breathing exercise can profoundly soothe and stabilise your emotional state, promoting emotional resilience. Incorporating these techniques into your daily routine ensures that your heart chakra remains balanced and your emotional health is fortified.

Empathy, Compassion and Emotional Balance

If you find yourself deeply affected by others' emotions, you may be an empath. While empathy allows for deep connection, maintaining clear boundaries through compassion is crucial for protecting your emotional well-being. This balance ensures that your heart chakra remains open and harmonious, aligning with the Law of Attraction to attract positive and supportive energies. When you feel drained by someone's energy, prioritise self-care. Acknowledge your achievements, offer yourself kindness, and then extend your compassion outward.

To maintain a balanced heart chakra and align with the Law of Attraction, it is essential to practice compassion towards yourself and others. When you feel drained by someone's energy, prioritise self-care. Acknowledge your achievements, offer yourself kindness, and then extend your compassion outward. Avoid immersing yourself in lower vibrational energies, such as emotional distress or discord, as these can diminish your own vibration. Instead, focus on maintaining a high vibrational state, which supports both your well-being and the well-being of those around you.

By upholding these practices, you ensure that your heart chakra remains open and balanced. This alignment helps you attract harmonious energies and experiences, in accordance with the Law of Attraction, enhancing your ability to manifest your desires. In doing so, you also become a beacon for others, teaching them how to stay balanced and cultivate their own high vibrational states.

Infinite wisdom now opens the gates to Divine healing, releasing all pain and expectations from the past, allowing me to move forward in serenity, under grace and in perfect ways.

Luella Goethals

True Story: The Gift

"The heart is the only broken instrument that works."
— UNKNOWN

Some experiences in life bring us to the edge of our understanding, challenging us to find strength and meaning in the midst of sorrow. This story touches on one such moment—a deeply personal encounter with loss and the unexpected paths it led to in the journey of healing and spiritual growth.

A pale Easter sun seeped shyly through the small window. The walls, stained with age, and the dull grey tiled floor seemed to absorb the faint smell of bleach lingering in the room. The only furnishing was the bed with adjustable bars. My husband sat close by, holding my hand. The nurse entered, cradling a bundle. As she lowered her arms, I held my breath. The head, much larger than I had anticipated, was attached to a perfect, small human body. This little one's heart had been beating until that final, violent spasm. Despite the medication, the contractions had started too early and never ceased.

I had envisioned the foetus as incomplete, with translucent, thin skin, particularly at five and a half months of gestation. As I released my breath, the melancholy that had gripped me seemed to seep out of my body, and the salt from my tears stung my skin. My mind wandered.

I recalled the birth of Louise Brown, the first "test-tube" baby, which had ignited ethical debates during my time at a Catholic secondary school. The concept was dismissed as morally unacceptable by many. Even then, I admired how scientific advances had brought joy to a couple after nine years of heartache. Little did I know that such breakthroughs would eventually become a significant part of my own journey.

After four years of fertility treatment, our journey culminated in loss. My grandmother described the miscarriage as 'God's will,' but I have since come to understand that the Universe operates in mysterious ways. Through my own pain and loss, I have learned spiritual lessons that I now share with you. We reside in a loving Universe where Divine Will aligns with grace and perfect order. Energy flows where attention goes. Instead of dwelling on loss, I faced this trial with courage, recognising that the soul I lost had a different path. As parents, we guide but do not own our children's souls. Their journey is their own, and our greatest gift to them is the freedom to choose and learn under grace and in perfect ways.

Despite this understanding, I could hear ancestral voices suggesting that I deserved this pain and loss due to past actions, interpreting it as God's punishment. Yet, I now know that we exist in a loving Universe. I learned not only from the pain of sorrow but also the importance of shifting focus, releasing pain, and embracing serenity on this journey. We did not officially declare the baby as our firstborn under French law, but I now acknowledge that child with a name.

By opening your heart chakra, you align yourself with the loving energy of the Universe, recognising that all unfolds in Divine Order. Accept yourself fully, balance giving and receiving, and focus on appreciation. This alignment elevates your vibrational frequency, attracting experiences and manifestations that resonate with your true desires. Through this journey, you honour the healing process and remain open to the abundance and grace that the Universe offers, knowing that everything unfolds perfectly and for your highest good (the greatest benefit for your overall well-being and growth).

And because I maintain my vibrational elevations, my dreams are fulfilled, under Grace and in perfect ways.

The Throat Chakra: Resonance of Expression

> "When words are both true and kind,
> they can change the world."
> — Anonymous

THE THROAT CHAKRA, KNOWN IN SANSKRIT AS VISHUDDHA, is the centre of communication, self-expression, and truth. Located at the throat and characterised by a deep blue hue, this chakra is connected to the element of sound. It governs our ability to express ourselves clearly and authentically. When balanced, the throat chakra allows us to speak our truth and listen deeply. The vibrations of your words resonate with the Law of Vibration, influencing the reality you create. When healing is needed, this chakra may appear dark or incomplete.

The Healing with Sound

Sounds are woven into the fabric of our lives. Water runs, doors click, cars beep, planes thunder, bees buzz, butterflies flutter, birds tweet, rain lashes, cannons boom in celebration, and horns once echoed ancient cries. This symphony of sounds surrounds us, shaping our experiences and emotions. Ancient civilisations understood the profound healing qualities encoded in sound, and they tapped into its energy through mantras, chants, hymns, and sacred incantations. Even the simple sound of 'Om,' considered the primordial sound of the Universe and believed to contain all other sounds, holds spiritual significance. The intentions embedded in these resonances were a crucial part of their healing power.

Music, with its universal language, transcends barriers of age, race, culture, and ability. Its melodies penetrate our senses, evoking memories and emotions, from joy to nostalgia. Beyond mere entertainment, sound is energy in motion, travelling effortlessly through the air. Surrounding air particles vibrate and collide, creating waves that ripple outwards. Harsh sounds bombard the atmosphere, while harmonious sounds form gentle undulations. Your words, like sound resonances, are powerful energies. Spiritual seers, past and present, remind us of the power of the spoken word. The vibrations of your words, coupled with your intentions and emotions, become potent transmissions.

As we explore the resonance of words, it is essential to recognise their historical and cultural significance. Across civilisations, words have been revered for their ability to shape reality and connect individuals to higher realms. Ancient Sanskrit mantras, for instance, were regarded as sacred vibrations capable of healing and transformation. These practices continue to inspire modern sound therapies and mantra meditations, where the healing power of sound is harnessed for overall well-being.

The Resonance of Your Words

Your words, like your thoughts and emotions, carry their own vibrations. These vibrations are captured by your throat chakra and resonate through your subconscious mind. When your spoken words are filled with complaints and criticism, they emit low, heavy vibrations, often attracting similar patterns into your life. Like attracts like. Pay attention to the nature of your speech. Are you engaging in self-criticism or disparaging others? Consider the clarity of your intentions. Are you guiding and teaching, or attempting to control and dictate, potentially infringing on free will?

Take time to reflect on your communication. Are you offering constructive feedback, or merely expressing frustration? Remember that others often mirror aspects of ourselves. Before addressing someone strongly, pause and reflect on your own words and intentions. For instance, if you need to discuss boundaries, first consider where you might have crossed your own limits. Then, speak with gentleness and honesty.

When discussing morals or values, recognise that there are many truths. Your perspective is one among many. Approach with compassion, ask thoughtful questions, and release your expectations, respecting the free will of others. Often, the messages we wish to convey carry lessons for ourselves as well.

The Law of Vibration responds to your thoughts, emotions, and words. Destructive speech creates destructive vibrations. Your role is not to control or instruct others but to express your truth in alignment with your soul's essence, allowing others the freedom to choose their path. Consider the change you wish to bring about and how you plan to achieve it. Are you aiming to effect this change by controlling others, or by letting your inner strength shine to enlighten them?

Speaking out is valuable, but always examine the intention behind your words. Strive to make the world a better place through the resonance of your words, acting as beacons of light. When judgment and criticism are voiced, they create disturbances. Approach moments of conflict or criticism as opportunities for both personal growth and evolution.

Evaluate the context of the feedback you give and receive. Recognise when you might have crossed boundaries, and use this awareness to guide your interactions. Break free from cycles of judgment and criticism. Focus on your own evolution and let your words and intentions inspire and uplift. Rise to this challenge.

The Power of Discretion

There is no need to talk about your projects prematurely. They are delicate, and once spoken about, can attract low-resonating energy, potentially hindering their growth. This energy can be either unintentional or intentional. The shoots of your projects are young and fragile, requiring your care. Premature exposure could prevent them from reaching their full potential. It is not a matter of secrecy but rather of focus. By discussing your projects with many people, you risk dispersing or thinning their energy. The more compact and focused your energy, the better—unless you are seeking advice in an environment where you truly value expertise and constructive feedback. Friends and family, while well-meaning, may offer feedback based on their own experiences and choices in life.

Understanding the laws of the Universe, it is wiser to keep your projects protected and only expose them once they are in full swing. Speak about yourself or your projects when they are fully realised and ready to be shared. Inspire others through your success.

To align with the Law of Attraction, remember that what you emit, you receive. If you are critical, your projects may be dampened by your own words. Many spiritual teachers remind us to use our words wisely, as they have the power to shape our reality.

Embracing Your Voice

If you find yourself struggling to express your thoughts or feeling a tightness in your throat, try writing about yourself and your truth. Take time to reflect on who you are, independent of family or societal expectations. What version of yourself would you like to project into the essence of the Universe or the quantum field (a theoretical space where all possibilities and potential outcomes exist before becoming a reality)? Hold this image in your mind, live it out, and bring it into your physical reality. This practice aligns with the Law of Attraction, helping you to create a better version of yourself.

Express your boundaries clearly when others overstep them. Also, learn to listen to the natural sounds around you—the wind, the sea, water, birds, cicadas, and traffic. Notice the synchronicities and messages they bring. This practice enhances your ability to both receive and emit vibrations, completing the circle of communication. Tune into your intuition of clairaudience (the ability to hear beyond ordinary sound). Appreciate the silence, the unspoken, and the songs of Nature.

And because I am aligned with my own song, all my needs are met. In the silence, I exchange noise for peace and am at peace. Under grace and in perfect ways.

True Story: Louise's Weekend

> "Words are, of course, the most powerful
> drug used by mankind."
> – RUDYARD KIPLING

Words carry weight, shaping not only our reality but also the realities of those we care about. In this story, we explore how one woman's well-intentioned words became entangled with her fears, revealing the deep connection between our speech and the energy we project into the world.

Louis's throat chakra was grey and choked with austere words.

"Louise," I almost whisper, sensing her constricted throat, "have you had any recent problems with your vocal cords or pain around your throat area?"

"Well, yes! After a weekend away with friends, talking entirely about my daughter, I returned feeling suffocated. The doctors at the emergency centre prescribed Ventolin to facilitate my breathing. But I'm just fine now." She stops abruptly.

Louise is a mature and cultivated lady. This is her lunch break, and she is taking the time to find practical solutions to a profound dilemma.

"Actually, I'm here to talk about my daughter." Her tone is clear and crisp.

"Yes, yes, I understand." With a wave of my hand, she begins talking about her daughter.

"I'm preoccupied with her well-being. She's just lounging around the house, doing nothing. She's not even making any effort to look for work. I've asked her to see a psychologist, but she's not willing." Louise pauses, seeking reassurance.

"Yes, absolutely, you're right to encourage your daughter to consult a doctor." Louise settles back into the armchair.

"But tell me," I ask, concerned about her throat chakra, "is there anything that is 'right' with your daughter?"

Louise looks at me blankly, surprised that she cannot instantly reply. She realizes she is placing considerable attention on her daughter's deficiencies. Her energy depletes when she emotionally discusses her daughter's inadequacies. Her daughter, moreover, becomes the monster she constantly projects through the power of her words.

Louise has been focusing all her energy and attention on her daughter, practically choking on her own words. In focusing on her daughter's lack of endeavours, she unconsciously describes herself; she has suspended her personal and professional projects. Healing comes through resonating at graceful undulations of thoughts, emotions, and intentions, inside-out. She intends for her daughter to heal but has not fully understood the fertility of the subconscious mind. Her words, thoughts, and emotions propagate and feed the gardens of her subconscious mind and her daughter's with irregularities.

Louise's thoughts require direction and purpose, firstly for the evolution of her soul. Through our discussions, she understands that taking care of her mind and soul builds her inner strength. Positive vibrations, inside-out, ripple outwards like a beacon of light, reaching the person she dearly loves—her daughter, who has her own learning, experiences, and journey in life.

Your words have an authority of their own. They shape your reality and the reality of those around you. Speak mindfully, with intention, and align your words with the Law of Vibration. By doing so, you create a harmonious resonance that attracts positivity and manifests your desires.

Because I maintain my verbal resonances of truth, I manifest my dreams under grace and in perfect ways.

The Third Eye: Resonance of Insight

"The eye sees only what the mind is prepared to comprehend."
– Henri Bergson

THE THIRD EYE CHAKRA, OR AJNA, IS OUR CENTRE FOR INTUITION, insight, and inner vision. Positioned between your physical eyes, this deep-indigo chakra connects you with the unseen forces of the Universe. Acting as a lens, it allows you to perceive beyond the physical realm, aligning your thoughts and vibrations with the reality you wish to create. A balanced third eye chakra naturally focuses your attention on your desired outcomes. When adjustments are needed, think of refining your "lens" as adjusting a ship's compass to ensure a steady course. Just as a ship must constantly correct its heading to navigate the ocean's changing conditions, fine-tuning your focus helps maintain clarity and direction throughout your journey. By aligning your thoughts and intentions with empowering and high-energy states, you enhance your connection with the Law of Attraction, allowing you to visualise and move towards your goals with greater precision and ease.

Activating Your Intuition

The third eye is a gateway to higher consciousness and deeper understanding. When activated, your intuition guides you to take inspired actions that feel effortless and natural. Trusting your intuition begins with setting and focusing on clear intentions.

From my many teachers, I have learned that setting clear intentions and writing them down is a powerful way to activate your intuition. The third eye chakra, located between your physical eyes, is stimulated when you are specific about your desires through consistent, written intentions. These intentions send signals to the Universe, creating synchronicities that align with your goals.

With clear intentions, subtle thoughts and feelings guide you into action. This guidance is your intuition at work. Support this process by keeping your intentions clear and allowing them to work in the background while you focus on other aspects of your day. Trust the process, let the Universe handle the "how," and follow the intuitive guidance you receive.

Enhance your intuition by setting clear intentions as if they are already becoming reality. Write down your desires and let your intuition lead you towards them. This practice sharpens your intuitive abilities and aligns your energy with your goals. When inspiration strikes, act with enthusiasm and ease, avoiding tasks that drain your energy. Embrace your intuitive guidance and enjoy being in the flow of life. This alignment with your inner guidance system helps you harness the Law of Attraction, bringing your desires into reality.

The Power of Inner Vision

The mind's eye, often referred to as the third eye, allows you to perceive beyond the physical world. This inner vision is a powerful tool for shaping your experiences. By vividly visualising what you want to achieve, you create a mental image that sets the stage for it to become reality. This process aligns with the Law of Attraction, suggesting that focusing on and believing in your goals will help them come to fruition.

Supporting this, NASA's research shows compelling evidence. Astronauts who mentally rehearsed each phase of a rocket launch displayed brain activity patterns nearly identical to those observed during actual simulations. This suggests that visualising an event can stimulate the brain similarly to real experiences, highlighting the power of mental imagery.

Moreover, athletes recovering from injuries use mental imagery to stay connected to their sport and accelerate their healing. Mental imagery is a recognised tool in sports psychology and rehabilitation, aiding in performance and recovery.

By training yourself to vividly imagine and feel what you desire, you impress these intentions upon the Universe. This "substance" of the Universe is flexible and responsive, shaped by your thoughts and feelings, provided your energy centres—your chakras—are aligned and flowing correctly. Focusing your mind and emotions on your goals helps bring them into tangible form.

Key Words for Momentum

Setting intentions through key words is a potent method for aligning with the vibrations of your desires. Choose three key words that resonate deeply with your goals. For example, if your aim is to achieve a healthier body, you might choose words like 'nourished,' 'energised,' and 'tonic' to align with your intention. Write these words down and allow them to guide your focus throughout the day.

These key words act as vibrational anchors in your mind, accumulating energy much like a snowball gaining mass as it rolls downhill. The Universe responds to these vibrations, aiding in the attraction of experiences and outcomes that match your intentions. By immersing yourself in the energy of these words, you cultivate a deep sense of connection with their essence. This practice aligns your mental and emotional states with your goals, ensuring that your thoughts and feelings are in harmony with your desired outcomes.

The Power of Synchronicities

The third eye chakra, often seen as the seat of intuition, opens you to the synchronicities of the Universe. When you align with your inner vision and intuitive insights, you may notice more synchronistic events occurring in your life. These subtle signs and coincidences often guide you toward your desires. Pay attention to these signs and be open to following their lead.

When you experience a sudden, compelling urge to take a particular action, view it as inspired guidance from the Universe. These impulses often signal that you are on the right path and moving towards your highest good. Embrace these moments with appreciation, trusting they are part of a supportive process. This flow of synchronicities exemplifies the Law of Attraction, where your focused thoughts and vibrations attract similar experiences.

The Power of Vibration and Attraction

The Law of Vibration asserts that everything in the Universe emits a distinct frequency. By aligning your thoughts, intentions, emotions with your desires, you elevate your own vibration, attracting corresponding experiences into your life. The third eye chakra helps you see beyond immediate appearances, connecting you with deeper vibrations that shape your reality. This alignment is essential to the Law of Attraction, which brings into your life what you consistently focus on and believe, as believing is seeing.

Trust your inner vision and intuition. Allow your third eye to guide you towards your dreams, recognising that what you envision can manifest in your physical reality. This trust amplifies your ability to attract what you desire, reinforcing the connection between your inner vision and the Law of Attraction.

By aligning with the unseen through the third eye chakra, you enhance your capacity to bring your desires into reality. Activating your intuition, using your mind's eye to visualise, and setting intentional key words helps you stay focused on your path.

This alignment supports both the Law of Vibration and the Law of Attraction, attracting positive experiences and fulfilling your dreams.

And because I trust my inner vision, all my desires are realised, under grace and in perfect ways.

True Story: Seeing Clearly

> "The real voyage of discovery consists not in seeking new landscapes, but in having new eyes."
> – MARCEL PROUST

Sometimes, the most profound changes in life occur when we least expect them. This story reflects on a moment that led me to question everything and ultimately set me on a path toward spiritual awakening and self-discovery.

The momentum of my wedding day gathered pace. My parents sent out invitations, and the arrangements for the ceremony were made, including the special permission granted to my late uncle, a Jesuit priest residing in Pune, India, to perform the ceremony. Catalogues for dresses arrived, the hall was hired, and the catering was organised.

One evening, after discussing the marital rites with the priests at our local church, their words stirred a deep, unfamiliar sensation within me. As I drove home, I moved more slowly and with intense concentration, focusing on changing gears, stopping at traffic lights, and navigating roundabouts. An inner struggle began as I grappled with a swelling emotion I could neither fully identify nor suppress. The streetlights flickered on, casting a dim glow over my path. Upon reaching the flat, I fumbled through my bag for the house keys and accidentally grabbed my spectacle case instead. The case felt heavy in my hand, and I saw my glasses inside it. In a panic, I blinked and looked around frantically. Glancing up at the bathroom window, I remembered my contact lenses were still sitting in their disinfectant solution. Desperation took over as

I screamed, hands clutching my eyes, yearning for clear vision. I felt an urgent need to express my fogginess.

"I cannot go through with the wedding. Something just doesn't feel right."

My mother glanced over, her face a mask of disbelief. Her hand waved dismissively. "What do you mean? It's too late, and all the invitations have been sent out." She turned her back on me and hurried out of the kitchen, into the narrow, dark landing, and up the stairs. I sat there, feeling lost and questioning what I had hoped to achieve by opening up to her.

With hunched shoulders, I walked out of the front door and stopped abruptly.

In my recurring dream, I run away effortlessly across the road, down the path, and into the park fields, running with such vigour and so far away, to some distant place.

Confiding in a loyal family friend and releasing those pent-up emotions brought a temporary solace.

The tune playing on the radio as I sat on the edge of the bed in my old room interrupted my quiet reflection. I looked in the mirror, noticing how the high heels lengthened my silhouette and the magnificent dress. I took a deep breath and straightened myself. That hot August afternoon, as I walked down the church aisle with my father, I vowed to myself that I would find clarity and get it right the second time around.

We moved to Brussels and divorced under Belgian law that January, just five months after marrying. As anticipated, my parents severed contact with me. The marriage was annulled three years later, which not only liberated me from religious constraints but also marked the beginning of my spiritual journey. My Third Eye began to open, enhancing my intuition and allowing me to connect with the unseen, including exploring the chakra system.

I am under grace and I am always at the right place, at the right time with the right people.

The Crown Chakra: Resonance of Divine Connection

> "The crown chakra is the portal to our highest self and the infinite consciousness that lies beyond."
> – Louise Hay

THE CROWN CHAKRA, THE SEVENTH CHAKRA, SERVES AS OUR gateway to Divine consciousness and spiritual connection. When open and aligned, it enables access to higher states of awareness and enlightenment. The colour associated with the crown chakra is often white, symbolising purity, clarity, and the boundless possibilities of the Universe. Located at the top of the head, this chakra acts as a bridge between our individual self and the cosmos, or infinite wisdom.

In relation to the Law of Attraction, the crown chakra plays a pivotal role in aligning your intentions with the greater cosmic forces. When this chakra is balanced, it allows you to connect with the universal flow of energy, enhancing your ability to manifest desires that are in harmony with your higher self. An open and aligned crown chakra helps you transcend personal limitations, attracting experiences that resonate with your spiritual path and the limitless potential of the Universe.

Empowering Through Guidance

There are many ways to assist others, whether through offering your time or resources. As the saying goes, "Give a person a fish, and you feed them for a day; teach them to fish, and you feed them for a lifetime." Rather than merely providing solutions, focus on empowering individuals to find their own answers.

This approach keeps their minds engaged and fosters growth. For instance, if you bake a cake for someone, include the recipe so they can recreate it themselves. When reviewing an employee's work, ask questions that guide them towards self-reliance. If a teenager needs help with an application, support them in understanding and completing it independently. Although it might be quicker to do tasks for others, true assistance lies in nurturing their independence and personal development. Our time here is precious, and helping others become self-reliant is an essential part of our spiritual journey and legacy.

Living the Golden Rule

This fundamental principle, often known as the Golden Rule, is sometimes overlooked amidst the demands of daily life. Reflect on how you wish to be treated—with respect, kindness, and understanding—and incorporate these qualities into your interactions. By doing so, you activate the Law of Attraction, allowing the energy you give to return to you, though not always from the same person. Although simple, this principle can be forgotten in the busyness of modern life.

There is no need to change your personality; instead, focus on integrating this quality into your daily interactions to foster harmony. Simple acts, such as offering a seat to someone, refraining from criticism, and showing compassion, can significantly enhance our relationships. By living according to this rule, we maintain spiritual alignment and authenticity, creating a ripple effect that inspires others to follow suit.

Be the Change you Seek

To inspire more kindness in the world, start by practising kindness yourself. If you seek greater peace, nurture it within yourself first. If improving the environment is important to you, lead by example and promote sustainability. While you cannot change others, you can transform yourself and inspire those around you. By embracing the changes you wish to see, you align with your true self and bring more light and authenticity into the world. You cannot control others, only yourself, your outlook, and your perspective. By becoming the change you want to see, you encourage transformation in others. Advocate for what you believe in through your actions rather than attempting to control or dominate, as such approaches can infringe upon free will.

These three practices—helping others, living by the Golden Rule, and being the change you wish to see—align with the Law of Vibration: what you give out will return to you, enriching your journey and deepening your connection to the Divine.

Infinite wisdom now opens the gates of Divine guidance, under grace and miraculous ways.

True Story: The Apple Tree

> "The greatest and most profound truths are often the simplest and the most accessible to the heart."
> – ANTOINE DE SAINT-EXUPÉRY

That day, in the quiet stillness of waiting, I unknowingly stepped into a deeper understanding of the complexities of life and the path toward healing that would shape my future.

I was excited. Mum was coming home that sunny afternoon. I supposed it was because my birthday was soon approaching. I missed Mum. I had overheard conversations that she was tired and needed to find recluse with nuns. All that morning, Dad patiently dealt with a five-year-old trailing him around our small home in North London: in the kitchen, through the dining room, and in and out of the garden. In the middle of the garden stood our apple tree, full of small green apples, patiently ripening in the August summer heat.

The big hand of the dining-room clock pointed straight up, while the smaller hand showed that it was five o'clock. My father told me to sit down, and the expectation of being a good girl and not drawing attention to myself compelled me to comply. Fiddling with my wiry hair and swinging my feet wildly while seated on the imitation black leather chair, I kept glancing at the clock and towards the front door that was clearly in my view.

I felt like a young person with large eyes absorbing the world, not always understanding what was going on, but expected to think like my parents in their mid-twenties. My mind wandered, as usual. I thought about the day Dad came to collect me from nursery school in Kenya, almost three years ago.

The metal gate behind us clicked open, startling the afternoon's routine. We instinctively turned around in unison. I gasped. My dad was strolling up towards us. He was wearing his baggy black trousers and white short-sleeved shirt, although he no longer went to work. What was he doing here at this time? Did he have to pick me up earlier than usual because Mum was busy again?

As he approached Mother Superior, he firmly extended his hand. The nun's hand, appearing from somewhere within the folds of her long white habit, looked pale and fragile enveloped in my father's mighty hand. Facing the group, Dad searched for my brown eyes amidst the thirty pairs belonging to the diverse group of playful three-year-olds. Meeting my wide eyes, he winked and smiled at me warmly. Then he abruptly turned to Sister. Holding up his camera, he pointed to the lens and a button. Dad took great care of his camera and loved to take the time to focus the lens before taking the photo. The camera was often dangling on his shoulder. I sighed. I assumed that Sister had never seen a camera simply because she had never seen a television.

Sister nodded, and as Dad walked briskly back towards the gates, she motioned us to gather around the jungle gym, explaining that it was for a photo shoot. She placed me first, standing in the hayed grass. Girls and boys soon stood on either side of me. Others settled cross-legged in front, while the remaining children sat high in the climbing frame behind. We looked straight ahead, obediently, eyes fixed on Dad, genuflected, the camera hiding his face. We were still and silent. My hair, absorbing the heat from the tropical rays, ordered my feet to fidget in the bald straw patch. The camera clicked; twice. Contented, Dad came swiftly to my side.

He took my small hand in his. Bending down, he whispered in my ear. "We're leaving Kenya, leaving Nairobi, for good. We're leaving this very evening. We're going far away - to Europe. We are going to fly on a Jumbo 747. You're not going to see your little friends again." He gently shook the camera before my eyes. "The photo is going to be your memory."

As we headed towards the gates, my feet froze. I looked back. The nun was holding the hands of a girl wearing a Short, purple-checked dress that exposed her thin and fragile legs. The girl jumped gingerly off the frame, landing on her feet, knees bent, smiling at her graceful jump. Perhaps this was the moment to say something to Sister.

"It's quite strange," Sister had said to my mother one morning when we had arrived late. "There are days when your daughter is bubbling and so willing to participate in our play rehearsal. But there are many days when she is totally silent, even sullen, refusing to speak. We cannot give her a major role in the play as she is so unpredictable."

On my silent days, Sister would kneel to my height. Holding my hands with her warmth and safety, she would look deep into my eyes, waiting for me to speak. My lips remained intuitively sealed, but through the windows of my soul, I knew she saw my trapped and nameless emotions. Her gentleness was inviting, but my alliance with my mother was stronger. Sister only sighed and let me be. I never did get the chance to tell that nun.

"It's time to go." My dad's voice made me look up at him. I nodded and turned away from the playing children. I tightened my grip on his hand.

The blue front door finally opened wide, bringing me back to the present moment. Dad appeared at the entrance from seemingly nowhere. He grabbed my mother's arm and marched her swiftly into the room. They brushed past an invisible me. An echo from Nairobi sounded warnings.

Dad stood Mum against the wall, at the far end of the room, but still within view of an unseen child. Mum's skin was a sickly yellow. She rushed a glance in my direction, and then, with eyes wide open, she followed every movement of the athletic body facing her. He was questioning her. Why had she run away? How had she taken all the money from the account? He could not pay the bills. This time he was shouting, and she was very, very silent.

He casually picked up the stick that lay on the dark printed carpet. It was the one that he had broken off earlier that morning from our apple tree. He extended his arm and held it up. Then he whipped it down. The skin on her arm slashed open. Blood oozed out, a brilliant red. Another crack resounded, and then another, shattering the innocence of the summer stillness.

Then the branch splintered across her arm. Silence and hesitation mingled together for a moment before he moved with agility into the garden for fresh branches. Her head was now bowed. Bound by the profound belief of duty, she made no effort to move. We were both trapped on the woman's ancestral island of silence, duty, and obedience.

On another occasion, while Mum was standing at the kitchen sink, a heated exchange sparked between her and Dad, who was standing further away at the door. A wooden stool with a blue seat suddenly hurtled past me and struck Mum on the head, causing blood to trickle down her face. At the hospital, I was asked the same questions by at least four different people. They noted down my answer that Mother had me memorise in the car. If I told the truth, I would be put into a foster home and would be breaking up the family. Did I really want that? No.

The lying began thereafter. Godfather called by the house unexpectedly the following day. There was no mention of what had transpired within the household. He just sensed that it would do me wonders to see an extraordinary film on the "big screen." He was right. *Mary Poppins* brought me immense joy, although Godfather fell asleep during the film.

The financial burdens grew heavier, and the undercurrent between Mum and Dad deepened. Mum and Dad held both day and evening jobs. Upon entering secondary school, wearing a second-hand uniform, unlike my peers, I consciously steered away from conversations concerning school trips that we could not afford. There were nights when sleep came to me briefly in the early hours of the morning. On such days, I took an exceptionally early bus to school. Sitting on the upper part of the red double-decker bus and gazing out of the window to see shopkeepers unlock the security blinds, people bustling in the streets as we rolled from one village into another provided me with a sense of anonymity.

One particular morning, I ambled along the cemented path of the calm school grounds. A muscle tightened somewhere around my left breast. My fingers fumbled to loosen the grey school scarf around my neck. Chatter filled my head. I was speaking to myself, as always. There is no need to go into the details. Just explain that you're feeling particularly tired today. Just say that you need to talk about some mixed-up feelings inside.

I paused before the old-oaken convent door. Breathing in deeply the damp autumnal air, I finally pushed open the heavy door. The high-ceiling entrance hall was now void of babbling girls. The chequered black-and-white tiles of the floor, faded and cracked, oozed coldness into the dim-lit air. I shuddered as an old memory resurfaced from just after our arrival from Kenya—chequered tiles and spilled soup, another moment of my awkwardness. I stopped in front of the familiar wooden staff-room door and knocked with resolution, A teacher opened the door. I swallowed hard. "Excuse me. I would like to talk with Miss Wright. I know she is usually available at this time in the morning."

"She's taken sick. She'll be in tomorrow. Can I help you?"

"No. No. It's fine."

I walked along the corridor and into my classroom, not knowing whether to be relieved or to cry with disappointment. Providence had intervened on behalf of Mother, who sternly forbade me from disclosing our family concerns.

"You wouldn't want to be responsible for breaking up the family and you and your brother going into different foster homes, would you?"

The subject was never mentioned again. I redirected my attention and worked diligently to gain admission to university, channelling my energy into academic success.

This is my parents' story, not mine. But what makes it part of my journey is the anger and pain that coursed through my veins and the steps I have taken over many years to find peace and healing. Even today, there are moments when old emotions resurface.

I am deeply grateful to my husband, children, brother, parents, mentors, clients, friends, students, and everyone who has crossed my path. Each of you has played a vital role in my journey, and you, in particular, have been a special teacher. Through our connections and experiences, I discovered my truth, my path, and solace in the spiritual laws I now share.

I once thought that others would provide the answers to my struggles, but I came to realise that true solutions come from within. This realisation led me to embrace the boundless power of the Universe and its timeless laws. The Universe brings people, places, and synchronicities into our lives to guide us. I've learned to focus on the resonances I wish to share from within.

Saint-Exupéry reminds us that the most profound truths are often simple and accessible to the heart. Wherever you are, whatever you're doing, if you experience unconditional love, even briefly, it is a remarkable feeling of existence. It is a resonance of grace and happiness that can transform you into a beacon of light.

Thank you for being a part of this journey and helping me open my heart to new depths of appreciation.

And so, Infinite Wisdom now opens the gates of Divine healing, releasing me from the past and guiding me forwards. I release any attachments to the past, understanding that in letting go, I create space for new and wonderful experiences to enter my life. Under grace and in perfect ways. It is done. It is done. It is done. Thank you. Thank you. Thank you.

Testimonials

Even after just a few hours with Luella, I've gained new perspectives and a deeper awareness of my personal journey. She's provided me with powerful tools for growth and transformation. I truly believe the Universe brought her into my life to help me embark on the transformative journey I've long desired. Now, I feel confident that I can achieve it. SV.

Luella is a highly intuitive coach with a professional approach in both her demeanour and communication. She effectively imparts her knowledge to her clients thanks to her excellent mastery of corporate communication and personnel management. Luella offers her clients solutions that they can easily integrate into their personal and professional lives. HC

I greatly appreciate working with Luella, who has a very human approach, naturally attentive to others. Her listening skills foster a gradual exchange, which sometimes prompts pauses for self-reflection. Rather than simply stating things, she helps you understand and encourages personal growth through your own insights. I understand that her goal is to help us grow personally and to become independent in our own development. Her work with energies is astonishingly precise, even to the point of awakening emotions within us. I sincerely thank her, first and foremost for her genuine kindness, and also for her professionalism. RF

Luella is outstanding—an extraordinary life coach. Throughout my ongoing journey with her guidance, I am experiencing profound *personal and professional change, navigating transformative growth and exceptional development.* NOH

I was able to resolve a painful and complex family situation with Luella's support during coaching sessions. Through her kindness and gentleness, Luella provided me with invaluable advice, support, and attentive listening. I appreciated the accuracy of her insights and her dedication, which restored my confidence to take action. Thank you from the bottom of my heart. FC

I met Luella 10 years ago when I was on the verge of burnout. Her kindness and positivity transformed my outlook on life. Throughout our sessions, I reconnected with my feelings, discovered my talents, and developed my skills. Thanks to her practical advice, I took action, gained confidence, and progressed professionally. I also learned to accept others as they are and to forgive. Today, I finally feel in the right place, fully in control of my life. MD

Luella provides a precise reading of energies and offers insightful advice on how to move forward in the direction you desire. Her guidance is both rare and invaluable. BC

References

- Abraham, H. (2006). Ask and It Is Given: Learning to Manifest Your Desires. Hay House.

- Beckwith, M. B. (2009). Spiritual Liberation: Fulfilling Your Soul's Potential. Atria Books.

- Chopra, D. (1994). The Seven Spiritual Laws of Success: A Practical Guide to the Fulfillment of Your Dreams. Amber-Allen Publishing.

- Cumming, J., & Williams, S. E. (2012). The Role of Imagery in Sport and Exercise. In S. D. Mellalieu & S. Hanton (Eds.), Contemporary Advances in Sport Psychology (pp. 129-149). Routledge.

- Dispenza, J. (2014). You Are the Placebo: Making Your Mind Matter. Hay House.

- Driskell, J. E., Copper, C., & Moran, A. (1994). Does Mental Practice Enhance Performance? Journal of Applied Psychology, 79(4), 481-492.

- Dyer, W. (2004). The Power of Intention: Learning to Co-Create Your World Your Way. Hay House.

- Hay, L. (1984). You Can Heal Your Life. Hay House.

- Hicks, E., & Hicks, J. (2004). The Astonishing Power of Emotions: Let Your Feelings Be Your Guide. Hay House.

- Judith, A. (2004). Wheels of Life: A User's Guide to the Chakra System. Inner Traditions.

- Kosslyn, S. M., Ganis, G., & Thompson, W. L. (2001). The Case for Mental Imagery. Oxford University Press.

- Lipton, B. H. (2005). The Biology of Belief: Unleashing the Power of Consciousness, Matter & Miracles. Hay House.

- Murphy, J. (1963). The Power of Your Subconscious Mind. HarperCollins.

- Myss, C. (1996). Anatomy of the Spirit: The Seven Stages of Power and Healing. Crown Publishing Group.

- Peale, N. V. (1952). The Power of Positive Thinking. Prentice Hall.

- Scovel Shinn, F. (1925). The Game of Life and How to Play It. DeVorss & Company.

- Three Initiates. (1908). The Kybalion: A Study of the Hermetic Philosophy of Ancient Egypt and Greece. Yogi Publication Society.

About the Author

Luella Goethals, a British Life and Career Coach with Indian origins, is dedicated to understanding and applying the mysterious forces of our Universe that shape our reality. With deep knowledge of metaphysical principles and ancient wisdom, she helps others navigate life's complexities with insight and grace.

Grounded in her own experiences of spiritual growth, Luella shares her knowledge with those seeking balance, inner peace, and a deeper sense of purpose. Her approach blends spiritual wisdom with practical guidance, empowering individuals to overcome challenges and embrace serenity.

Luella's mission is to help others turn life's challenges into opportunities for personal growth. She believes in evolving into a more authentic and serene version of oneself. Through her guidance, she inspires others to approach their journey with renewed energy and confidence.